Spend Green and Save The World

Tackling Climate Change Through The Consumer-Led Movement

Liz Christou

Oakamoor
Publishing

Published in 2021 by Oakamoor Publishing.

ISBN: 9781910773789

Illustrations: Hernán Parente [heparente@gmail.com]

For Athan

Acknowledgements

To my friends and family who have supported me with my efforts to write this book and replied to my constant requests for feedback. I thank you for this, and for supporting all my endeavours. I am blessed with the feeling of deep gratitude every day, because you are in my life.

About the Author

Based in the United Kingdom, Liz Christou has been solving complex problems for one of the world's leading brands since 2014. Before completing her MBA, and being selected for the esteemed graduate programme, she explored careers as diverse as city bar and restaurant manager and police officer, after leaving school at 16.

Liz has never followed the obvious path, or been put off by challenges that others have suggested are too difficult. Her journey to writing this book has been no different. She combined her educational and career experience with her aptitude for taking on tough challenges, and life-long love of the natural world, to tackle one of the biggest challenges we're facing today – climate change. Specifically, what we as individuals can do about it with the potential to make a real impact.

Liz used her experience in problem-solving – namely principles from Lean Thinking and Agile – to decrease her carbon footprint to 3 tonnes per year, from the UK average of around 8 tonnes, within 5 years, whilst having a positive effect on her wellbeing. Now, she has translated the journey into the motivational techniques and practical suggestions in this book, so that you can do the same.

Table of Contents

1 | Understanding the Problem and Being the Solution

Section 1 – Our climate problems

The problem, part 1 – our inner struggle

If you've picked up this book, you already know that there's a tsunami coming at us in the form of climate change. It's threatening to wash over the face of the globe and leave nothing, and no one, untouched.

If you're like most people who are engaged in the climate change issue, you know – on some level – that we all have a responsibility to avert this disaster, and you want to participate. You want to live in a way that respects the severity of the issue, and which demonstrates a suitable response. It's that you just don't know where, or how, to begin.

Whilst researching this book, I've come across many concepts and theories. Books, organisations, and charities talk mainly about political change, with only a small number giving guidance on what individuals can do. I've not seen anything that gives practical guidance on making both manageable and realistic lifestyle changes; all the while, juggling other priorities in our busy lives.

We need practical solutions that can pull us back from the brink of catastrophic climate change. But we also need direction, a method, and a goal to reach the lifestyle changes required. The purpose of this book is to help you navigate to the lifestyle changes with the biggest impact. And, to successfully reach the goal that can really make a difference.

The problem, part 2 – our external surroundings

The aim of this book isn't to go into detail about what the climate models predict for our future. There are lots of papers, books, and articles out there – not to mention YouTube videos from climate scientists – who will tell you first-hand what we have in store.

However, the stakes are so hard to swallow that it is worth reiterating *what is at risk*; because we want the reality of the situation to be at the forefront of our minds. It will prevent us from mentally logging it as something we *ought* to give more thought to… when we have more time.

There is no more putting this off. The outcome depends on *what* we do *now*.

Climate models show that by the end of this century, our current trajectory of greenhouse gas emissions will have warmed the Earth's climate by 3-4°C. This is the overall global average temperature, not whether a particular place is a bit cooler or warmer on a certain day. This means that we are moving the Earth out of the stable climate of the Holocene period, (experienced for the last 12,000 years, and which has allowed civilisation to thrive), into something more unpredictable and much less tolerable.

"From shifting weather patterns that threaten food production, to rising sea levels that increase the risk of catastrophic flooding, the impacts of climate change are global in scope and unprecedented in scale."[1]

The UK is thought to be one of the lesser affected areas. But on our current path, no country is predicted to bear a resemblance to today. More action is needed to prevent flooding, droughts, and "significant threats to our natural capital and the goods and services it provides, from timber, food and clean water to pollination, carbon storage and the cultural benefits of landscapes and wildlife."[2]

Let's be clear, we're talking about a struggle for clean water and food, and this type of struggle isn't pretty. People have no choice but to go into survival mode, which leads to wars, violence, and crime. We're seeing these impacts around the world already. George Monbiot explains that the effects are already visible, "In Mozambique, Zimbabwe and Malawi, devastated by Cyclone Idai, in Syria, Libya and Yemen, where climate chaos has contributed to civil war, in Guatemala, Honduras and El Salvador, where crop failure, drought and the collapse of fisheries have driven people from their homes."[3]

No one knows exactly how bad it could actually get. The climate models give us an idea, and we can hope that they're overestimating future impacts.

However, as we're dealing with something completely unprecedented, they could be *underestimating* the scale and timelines.

Melting ice caps, sea-levels rising, ocean acidification, and extreme weather events are already proving worse than previously predicted.

The problem, part 3 – our 'big-picture' issue

Climate change is waking us up to the huge void that lies between modern society and our natural world. It's a void between our behaviours and what is best for the planet, and it has a profound impact on our wellbeing.

Our lifestyles centre around one thing – *consuming*. Unsurprisingly, those of us who have taken a step back to think about that – and even the ones who haven't – realise, on some level, that this should not be our primary function as human beings. In fact, it's largely unfulfilling and causes immense environmental destruction (including climate change). Overconsumption is distancing us from a life that is connected to nature.

But would we really be better off if we lived in a way that respected our natural habitat? It sounds a bit 'fluffy' to many (or even most people) because we tend to accept the world from our current view, rather than challenge it.

The answer, though, is a resounding *yes*! It's becoming harder to ignore that our distance from – and lack of respect for – nature has come back to bite us. It's now all too evident, from climate change to rising mental health issues to the diseases caused by the food we eat.

We inherently need to be close to nature to fulfil our wellbeing needs, which we all know go well beyond the financial. But the only measure of success in our society is making more money to buy more stuff. So, we keep doing it, despite the negative effect it's having on us and our collective home. Getting richer and buying things will never fulfil us.

Before we delve into how and what we can do about it, it's important that we understand the root of all these issues. We need to know and understand the underlying cause of overconsumption which is creating climate change. And why we feel a block to taking individual action. Then we'll have an awareness that will empower us to put the right solutions in place.

Section 2 - The real root cause and what it has to do with us

"If I had one hour to save the world, I would spend 55 minutes defining the problem and only five minutes finding the solution." Albert Einstein

The following chapters in this book focus on how we, as individuals, can reduce our carbon footprint. They offer a way to do this collectively so we can begin to turn the dial on climate change in our favour. But before we can do this successfully, we should explore what's been blocking us so far.

We need to discover why we've cooked up a society that is not only causing climate change but which tolerates growing social injustice and is negatively impacting our wellbeing. As many writers, including Naomi Klein, and organisations like Extinction Rebellion and Greenpeace, attest, these aspects are deeply interwoven. This means that when we understand the underlying cause of climate change, the changes we make don't just tackle one of these issues, they can have an impact on all of them.

Failing government

New governmental policies on climate are a must, and will undoubtedly have a huge impact on climate change issues. Take charging for plastic shopping bags, for example. It's a tiny 'windbreak' in the 'environmental hurricane', but charging a measly 5p per bag in the UK has created a mindshift that transcends money. Plastic bag usage has decreased by 83 percent since 2015, that's 6 billion fewer bags a year!

Imagine if policymakers truly pulled their thinking out of the bag and backed up their Paris Agreement pledges with real action.

Big changes would happen.

The trouble is that they can only do so much while operating within the current economic system that relies on constant growth. Although, politically, we are operating within a democracy as well as a capitalist system, the fabric of all parts of our society – cultural/socio-economic and political – are completely intertwined. This makes government seemingly *as* subservient to the big players in business as 'the people'. And big business lobbies government relentlessly for:

- Continual increases in general consumption.
- Fossil fuel to be the main energy source for a growing population.
- No restrictions on intensive farming or deforestation.

These things are at the heart of rising greenhouse gas (GHG) emissions.

So what about big business?

Our system has evolved to focus primarily on making a profit in the short term. Asking CEOs for environmental protection to be their primary consideration is asking them to measure themselves by different standards of success entirely. That's unlikely to happen in the timescales needed while we're using a system that revolves around economic growth. Economic growth, in reality, equates to ever-growing consumption.

Within the consumer goods sector, the Consumer Goods Forum is a CEO-led organisation that aims to help the world's retailers become sustainable. Their initiatives aim to lower retailers' impacts on climate in line with the Paris Agreement. This sounds fantastic. But within our system, neither this forum nor the Paris Agreement can advocate what's actually required to avert climate disaster – *less consumption.*

It seems, then, that it's the system we're operating within that is blocking both governments and big business from making the changes required to avert climate change.

So:

- Is it the system that's responsible?
- Will changing the system resolve it?

Many left-wing thought leaders will tell you yes, but let's explore a little bit more to understand if the system is the real villain here. To do that, let's take a look (a very dumbed-down version) into the human brain. After all, our brains have created the notion of governments, businesses, and the system within which they operate.

The three main parts to our brains centre around our:

1.　　　Survival instincts.
2.　　　Emotions.
3.　　　Higher thinking.

The higher thinking part of our brain evolved last, and is what separates us from the apes and other animals.

Higher thinking is what makes us human.

It allows us to analyse our animalistic, instinctual behaviour – that comes from the more primal parts of our brain that we'd rather not act on – and move away from it.

Why is this relevant here?

Because it's the *less desirable traits* we display that are turning the cracks in the system into major flaws, and which are causing the climate change issue. Capitalism's original intent was about growing the economy to maximise *wealth for all*, but our animalistic characteristics have turned this into maximising profit/benefit for the individual at any cost.

This greedy, selfish behaviour – which the system perpetuates – comes from the primal parts of our brain that centre around survival and emotional thinking. Capitalism doesn't allow lots of time for higher thinking; above and beyond our primal and emotional impulses.

When we're thinking in our right (higher thinking) minds, we all want to be part of a society that considers wellbeing the most important goal. Looking after ourselves includes looking after our environment – the collective house in which we live. It should go without saying that our wellbeing is intrinsically linked to the health of the planet. Unfortunately, this seems to need pointing out too often.

The behaviours that capitalism perpetuates, in the name of looking after number one, are actually in direct opposition to a huge part of our wellbeing, which is all about taking care of each other. Yes, financial wellbeing is important as everyone deserves to be comfortable, but the focus on purely financial wealth is becoming our downfall.

We need to use the higher thinking brain to improve and change the system, in order to focus on our wellbeing and that of the planet. But, the system doesn't encourage us to use the higher thinking part of our brain. So the real root cause of climate change is a self-perpetuating cycle of the capitalist system, and our lesser nature.

We're stuck in a Capitalism Catch-22

The Capitalism Catch-22 or CC22 as I'll call it for ease, makes us forget about what's really important. It's as if buying more and more things will somehow quell our built-in urge to connect with other people and our natural environment. Unfortunately, to an extent it has. But it can't kill it completely because we need these connections for our wellbeing, far more than we need an extra car or holiday.

Sadly, modern, consumption-gone-mad capitalism has become so normal to us that, for many people, it seems crazy to question or attempt to change it. Even in light of the inequality and planetary destruction it's causing. It's so ingrained that it has become a way of life and a way of thinking. It keeps us so busy on the treadmill of more money-making for personal gain if you already have some, or for survival if you don't, that there's little time or impetus to think up something different.

THE CAPITALIST CATCH-22

The CC22 has a lot to answer for. From the polluted air we breathe to our unhealthy diets, to the lack of peace of mind because we don't spend time in nature, to looking after number one instead of feeling a sense of community. To feel contented and live a happy, healthy life, we must understand that we're caught in the CC22 trap. Then we can see that taking care of our natural environment for the greater good is ultimately the best thing we can do for ourselves.

Section 3 – The Consumer-Led Movement

Good buy to climate change

Not using our higher brain capacity has made us forget that we are a part of nature, and that we rely on it to nourish us in every possible way.

Governments and business can make all the changes they like, but they will never be enough to sustain the planet if our personal attitudes – to respect nature for how important and powerful it is – don't change. If we continue to disrespect our natural environment while governments put environmental repair policies in place, governments will always have to fix the issues caused by billions of individuals' disregard for the natural world.

Thus, nature will continue to be plundered until the ecosystems are completely destroyed, and there is truly nothing left. I can't help thinking that even if humanity found a way to survive under these circumstances – which we're well on our way toward – it will be a stark and depressing existence.

The good news is that in a system where businesses fight for consumers' attention to survive, and consumers and businesses influence governments, it is the consumer that holds all the power…

"While consumers have always had the ability to vote with their feet, or with their wallets, they now have more power to influence not only what they buy, but also what others buy. Empowered by social networks and digital devices, consumers are increasingly dictating when, where and how they engage with brands. They have become both critics and creators… expecting to be given the opportunity to shape the products and services they consume."[4]

We can move towards societal wellbeing and minimise climate change by:

- Supporting organisations that create a green economy and provide products and services that minimise our impact, and even more importantly…

- ...Stop paying for products and services we don't really need that add no real value to our lives and which are environmentally damaging.

Individual actions add up

Analysis published in the Journal of Industrial Ecology[5] showed that consumers are responsible for *more than 60 percent* of the globe's greenhouse gas emissions – a figure which is made up from the direct use of carbon like heating our homes, and indirect uses, such as what we buy.

The research states that households have a relatively large degree of control over their consumption, but they often lack accurate and actionable information on how to improve their own environmental performance.

That means it's *our* choices that have the most significant impact on climate change. China might be the biggest emissions contributor as a country, but not because the people there are living the lives of Riley on some fossil fuel-burning, luxury lifestyle mission to destroy the planet. It's because they're making all the crap that we choose to buy!

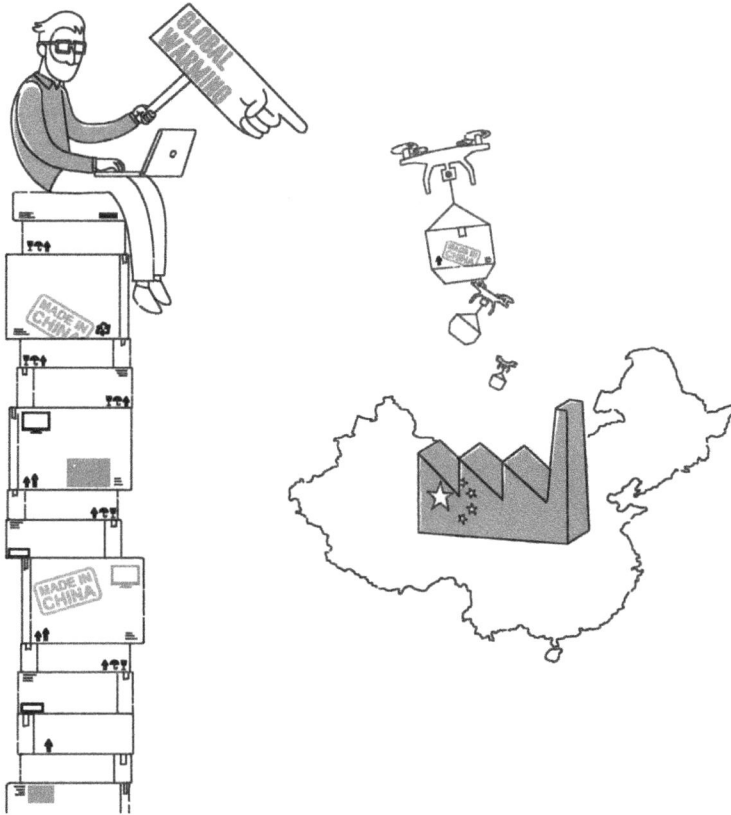

One reason we haven't acted on this as a society so far, is because there's no tool that brings us together to make our united actions count. It's hard to believe that you're making an impact on climate change and influencing society when you think you're acting alone.

(A minority of) People Power

Making cultural/socio-economic and political shifts happen is quicker and more achievable than you might think. You don't need every person to adopt a mindset change. A small percentage is all it takes to bring the rest of the community – whether that be local or national – along with them.

The anthropologist Margaret Mead, recipient of the Planetary Citizen of the Year Award in 1978, is quoted as saying, "Never doubt that a small group of thoughtful, committed citizens can change the world: indeed, it's the only thing that ever has."[6]

Recent research by Harvard Professor Erica Chenoweth puts some numbers to this notion. Her research found that for civil resistance to be successful, it takes 3.5 percent of the population to actively participate in a campaign.[7] And it's not just talking about taking to the streets for protests.

One example is the consumer boycotts in apartheid-era South Africa in which many black citizens refused to buy products from companies with white owners.[8] The result was an economic crisis among the country's white elite that contributed to the end of segregation in the early 1990s.

The recent massive turnouts at non-violent climate protests around the world show that millions of people are concerned enough about the crisis to take to the streets. Hundreds of thousands of people were, and still are, involved in several countries, including the UK, where 3.5 percent of the population is 2.3 million people. If hundreds of thousands are taking to the streets, then imagine how many of us – at home – also relate to the cause. A Yale University study in 2019 indicates it is a hell of a lot more than 3.5 percent. In fact, their study on climate change communication found that 72 percent of Americans described climate change as personally important to them.[9]

Enough of us are ready to put a plan into action, and for it to actually work. Read on to find out how to be one of the people making change happen.

And for the naysayers out there? Well, the wheels of change are already in motion. Take our mindset about food, for example. It has already made a massive shift in the last few years for environmental and ethical reasons, completely irrespective of government legislation. As a Forbes article detailing meat consumption explains, "Meat substitutes were projected to account for less than $2 billion of the projected total [retail sales of meat, poultry, and meat substitutes]. Wrong. The actual number was $4.63 billion in 2019. [Studies have underestimated] just how quickly meat substitutes are rising."[10]

When we get together, even inadvertently, we can, and do, make a huge difference!

Let's come together – right now

Today, there are 7.5 billion people, many of whom are thoughtlessly using the world's resources. At the same point, we also – as never before – now have the means to work together across the planet. Doing this will encourage the changes needed for a positive future without the threat of climate change, despite the lack of action from our so-called leaders. We can grow a new kind of movement. The Consumer-Led Movement.

CONSUMER-LED MOVEMENT
YOUR CLIMATE ACTION COUNTS

The information age

The web has evolved so rapidly in recent years that we've only just begun to catch on to its power as a tool to connect individuals and instantly share a world of information. This ability is now 'placed at everyone's fingertips'. We have seen connected movements, like Extinction Rebellion, where propelled action from people across the world has occurred within weeks.

Don't underestimate your power as an individual in this new age of sharing. The way positive mindshift changes are made, on a large scale, is when individual people come together to make a stand. People make a stand because they believe it is the right thing to do, regardless of what others are doing. They know acting on *what they believe in* is transcendent in nature. It will bring a contentment found only in aspiring to be something bigger than themselves.

As Desmond Tutu put it, "Do your little bit of good where you are, it's those little bits of good, put together, that overwhelm the world."

Key takeaways from this chapter

- Climate change is a dark cloud hanging over our future. How badly it affects our quality of life depends on what we choose to do now.
- The root cause of climate change is the Capitalism Catch-22. The system's reliance on consumption growth, and its look-after-number-one ethos, are perpetuated by our less-than-human traits, and vice versa. This makes it very difficult to break out of.
- Governments and big business are not able to take the action necessary to prevent climate change because they're also stuck in the CC22.
- We are all responsible for the issue, which is why we need to look at our individual lifestyles to effectively prevent climate change and further environmental damage.
- Understanding the CC22 allows us to think differently about what we spend our money on. Individually, this can help improve our wellbeing and reduce our carbon footprint. Collectively, this allows

us to reduce emissions to help prevent climate change and shape the system to serve us better in the future.

- There are more than enough people willing to play their part that we can create a shift in culture and positively impact climate change. The rest of the book provides the means and the motivation to get started.

2 | How to Navigate and Apply this Book Successfully

A new approach

So far, we've lacked a realistic approach for individuals to take action on climate change. We need to make improvements in our everyday lives that don't involve going to live in a treehouse, or weaving socks at home, else the prevention of more human-made climate change won't happen.

Before we get started, here's a brief overview of the method I've used to successfully reduce my carbon footprint without sudden, radical lifestyle changes. And, how it can make the journey not only possible but rewarding.

Lean and Agile are problem-solving approaches that are core to the success of many progressive businesses. The Lean/Agile approach (as I'll call it for ease) focuses on learning from our inevitable mistakes in order to keep incrementally improving. This is opposed to coming up with a perfectly thought-out plan of big changes and trying to roll them out in one go. This old-fashioned approach invariably doesn't go to plan and usually causes more problems.

So, if you're not quite ready to live caveman-style, but you do want to make positive and rapid improvements to avoid catastrophic climate change, then using a Lean/Agile approach to achieve your goals is the perfect tool. There's more detail on how to be successful using this approach in each chapter and also in the Appendix.

Here, let's touch on what is vital to succeeding, but what *most* books that give guidance on individual action lack…

Goals and measures

You have to know your individual goals and success measures for the improvements you make. Because, only then, you'll know *when* you've reached a lifestyle that is part of the solution rather than the problem.

1. Primary goal

Our primary goal is to reduce our individual carbon footprint to a level that helps us avoid global catastrophic climate change. It's recognised that to reduce the risks of extreme weather events and poverty, for hundreds of millions of people, we need to keep global warming below 1.5°C of pre-industrial levels.

The success measure

We need to put a measure in place to make sure we achieve the goal. Why bother putting effort into something if we don't know that it's bringing success, (i.e., getting us towards achieving our goal as fast as possible)?

Currently, as no one knows what success actually is, they can't prioritise. This has led to people focussing on the 'wrong' things or justifying doing very little as being enough.

The Intergovernmental Panel on Climate Change (IPCC) research[11] suggests that in order to have a good chance of keeping global warming within 1.5° C, we need to reduce our emissions globally by about 45% by 2030.

And reach net-zero by 2050.

The global average carbon footprint, per person, is about 4.3 tonnes per year. But this global average doesn't tell us anything about our individual contribution. Or, that those of us in the developed world are responsible for much higher emissions and have largely caused the problem with our lifestyles. For example, the average footprint in the UK is 8.46 tonnes per year, and in the US it is 17.75 tonnes. (This is the 'consumption-based' figure that places the emissions on the country that purchases the products rather than the one that produces them.)[12]

The global average also doesn't delve into issues around expectations that struggling economies – with low footprints – lower theirs. Hopefully, most of you reading this are reasonable enough to put the expectation of change onto the economies that have caused the issue. I realise that some of this is contentious. That said, endlessly debating the variables delays immediate action, and there is no argument that *immediate action is vital* if we are to prevent catastrophic warming.

So, in the target I'm proposing, I'm taking the IPCC recommendations into account. But I'm also respecting that responsibility lies far more heavily with the people with higher-than-average carbon footprints. The ones who have done more to create the problem than people in the poorest nations. (That's you and me, folks.) So, our key success measure will be tracking our personal carbon footprint, and bringing it down to:

<u>3 tonnes</u> of CO2e per year[13]

To know where you are against achieving the 3-tonne target, you'll need to keep an eye on your personal carbon footprint. To do this, we're using a carbon footprint calculator from the Global Footprint Network. It's available online at http://www.footprintcalculator.org/

Go there to find out your current footprint now!

For simplicity, and for the purposes of this book, we are only interested in the carbon footprint reading on the calculator, not the ecological one.

2. Secondary goal

In chapter 1, we talked a lot about our collective wellbeing suffering at the hands of the CC22; our collective focus on increasing wealth instead of connecting with nature is destroying our natural home. And focussing on increasing wealth means less time connecting with each other. The practical solutions in this book all reduce your carbon footprint, but because they are based on an understanding of the CC22, it's my experience that they have other benefits.

I feel they can also inspire a deeper connection between our behaviours and their effects on the natural world. And because they help us focus on the greater good – rather than materialistic pursuits – our personal wellbeing can heighten.

Scientifically speaking, there is no single definition of wellbeing, but some of the key areas incorporate:

- Physical and emotional components.
- Purposefulness.
- Achievement.

Based on these areas of wellbeing, I'm proposing the following goals to aim for, as individuals, with the improvements we put in place:

- A closer relationship with your natural surroundings.
- A sense of creating a positive future for your younger loved ones and yourself.
- Gaining some knowledge and skills to help fight climate change.

The success measure

You possibly won't feel the need to measure your wellbeing as it's something that you'll just sense and that will be enough. But, if you wanted to, you could give yourself a mark out of 10 for where you feel your wellbeing is currently at (against the goals set out above).

You can re-score yourself every time you reduce your carbon footprint by an amount of your choice. By reaching a lifestyle that avoids catastrophic warming (i.e., reducing your CO2e to 3 tonnes per year), the aim is to get your wellbeing score up to a 9 or a 10.

3. Stretch goal

The third goal is to be part of the Consumer-Led Movement. The aim is to reduce our *collective* footprint and create cultural change. Cultural change influences governments and businesses. The Consumer-Led Movement aims to shift culture towards conscious consumerism, to influence governments and businesses to create policies and practices that help us avoid catastrophic warming. We'll delve into how to achieve this in the following chapters.

Measuring the stretch goal

By being part of the Consumer-Led Movement, we declare that we're spending less on the things we don't need, and spending more wisely on the things we do. As proponents of the movement, we can begin to track the impact this is having on culture, businesses, and government policy.

Timelines

We're talking about pending human and environmental catastrophe. No one really knows the full extent of it, but the end of civilisation as we know it is predicted as highly likely if we don't take radical action within the next few years. With this in mind, as well as how long it has manageably taken me to reduce my footprint to 3 tonnes per year…

The set timeline to achieve our goals is 5 years

Key takeaways from this chapter

- When problem-solving, having a goal to aim towards, and a measure to judge success, are crucial. We, as individuals, have been lacking these so far for tackling climate change.
- We're aiming to keep global warming within 1.5°C to minimise the negative impacts of climate change. There is no exact measure for individuals to achieve this because of many fluctuating factors. That said, by collectively reducing our carbon footprint to 3 tonnes per year in developed nations, we can have a dramatic impact.
- We can also increase our wellbeing in the process:
  - We'll be escaping the trappings of the current system which focusses our thinking on materialistic pursuits.
  - We'll get a sense that we're doing something for the greater good by creating a positive future for ourselves and those we love.
  - We'll gain some skills and knowledge to tackle climate change and feel a deeper connection with our natural home.
- The stretch goal of being part of the Consumer-led Movement is also to impact culture and subsequently encourage environmental business practices and government policy.

- The timeline we're aiming for is to reach the 3-tonne carbon footprint goal within 5 years.

3 | Consumer Products

Overconsumption

Unless you're a bit of a hoarder, you can probably look around your home and find many pointless possessions that you've bought or been given. If you can't see them, then I'd bet money the last time you moved home, you were amazed at how much tat came out of the cupboards. It probably filled up boxes and boxes, and some of them were so full of who-knows-what that you haven't bothered to unpack them since.

We've all been there, rushing around the January sales to snap up any 'bargains' after we've just virtually bankrupted ourselves on Christmas. The sad thing is that the consumer-driven lifestyle we've grown accustomed to, has us convinced that we're doing ourselves a favour.

It's worth taking a minute to think about what 'consumer-driven lifestyle' actually means. We hear the word consumer all the time; it's completely normal to be called a consumer on a daily basis. We don't even think twice about it. Within our current system, the expectation is that to consume is our primary function. As a result – whether it be clothes, gadgets, furniture or TV – we're constantly consuming something.

This does huge damage to the climate due to all the greenhouse gasses emitted during the design, production, transportation, storage, retail, use, and finally disposal of the products. A lot of which don't add any real value to our lives. The Consumer Goods Forum[14] notes that the consumer goods sector is accountable for a whopping 60 percent of global emissions (this includes food, which we'll cover in a separate chapter).

For more guidance on how to get out of the consumer mindset, to benefit the climate and ourselves, read the breakout box. To start making change happen, head to the practical solutions.

Why do we buy, buy, buy?

Since the advent of capitalism, the system has evolved to perpetuate our innate thirst for more. This is done by constantly reinforcing the message that we need material things to rise up the status ranks. And that rising up the status ranks is a highly desirable thing to do.

I think most of us have the feeling that advertisers turn our psychology against us. They make us feel lacking if we don't have something, and it preys on our insecurities. What I didn't realise, until I recently watched a TED talk by Kate Raworth, is that the process is more deliberate than we realise.[15] She thinks we're socially addicted to growth and that it's due to consumer propaganda created by Sigmund Freud's nephew, Edward Bernays. In her words, Bernays "realised that his uncle's psychotherapy could be turned into very lucrative retail therapy, if we could be convinced to believe that we transform ourselves every time we buy something more."

Perhaps you accept that advertising has conspiratorial roots, perhaps not. But it's undeniable that we're constantly being influenced to believe that we'll be more desirable, interesting, or happy, if we buy things. And our primal, competitive, insecure side can't resist that message.

However it originated, this pattern of behaviour is now encouraged (to put it mildly) by those who benefit most from it. In the US, in the 70s, the number of ads the average person saw every day was around 500. In 2017, marketing experts estimate the figure to be between 4,000 and 10,000.[16]

Can we buy happiness?

We're definitely told to think that *what we buy* defines who we are. I have to say that I kind of agree, only not in the way the adverts suggest.

Ladies, it's clear that buying Tampax tampons isn't going to make you want to go rollerblading in white hot pants. And I'm sorry to break it to any teenage boys reading this, but spraying some Lynx on yourself isn't going to make a hundred semi-naked girls run at you.

We all know this, but thousands of adverts chipping away at our brain every single day takes its toll.

We end up actually believing that wearing a French Connection top makes us a little bit cooler, all because they put some slick adverts on TV and in glossy magazines, and *charge* you enough to believe they're worth more. They've managed to convince us that wearing a label with a four-letter swear word – spelt wrongly – makes us a more worthwhile person. It's fairly obvious that we're all being done. Or as Macklemore puts it, paying $50 for a t-shirt is "getting tricked by bizzzness".

In 'Prosperity without Growth', Tim Jackson puts it like this, "We're certainly not the first society to endow mere stuff with symbolic meaning. But we are the first to hand over so much of our social and psychological functioning to materialistic pursuits." For me, this quote really brought home the sad reality of our situation. We like to think of our society as being more advanced than any in history. But in taking a step back to look at ourselves, we see that we have our priorities all messed up to the detriment of our personal wellbeing and that of our collective home.

Despite all the attempts at brainwashing, we can't escape the problem that shopping is ultimately unfulfilling. Consuming is a distraction from the general discontentment that we feel because we're not accessing the higher part of our brains; the part that wants to create rather than consume. The part that feels energised by learning and doing new things, by being outdoors and nurturing relationships. No wonder so many people today are feeling depressed, and why practices like yoga and meditation have become popular. We just want some space to breathe, or attempt to minimise the incessant internal dialogue brought on – at least in part – by marketing campaigns that are designed to make us feel lacking if we don't buy their products.

It isn't just common sense that tells us this. According to the American Psychological Association, consumption "can promote unhappiness because it takes time away from the things that can nurture happiness, including relationships with family and friends."[17] And, despite becoming better off in the last few decades, our wellbeing has not increased.

Be the change and change what you buy

"Knowledge is Power" as Francis Bacon said. Ironically, he lacked much of it at the time; he thought that people can, and should, control nature. By understanding the root cause of the climate issue – 'The Capitalism

Catch-22' – we're aware that the behaviour the system perpetuates isn't coming from the higher brain. So we can *and should* analyse it before acting on it.

If we want to avoid catastrophic climate change and have a good chance to improve our wellbeing in the process, we can use that knowledge to make sure our buying choices define who we are. But not in a spoon-fed, superficial way. Instead, we can make conscious choices about what to buy, and what not to.

And, excitingly, when we make conscious choices about what to spend money on, these things can represent what we stand for and demonstrate our true values. When others notice what we do and don't purchase, it exhibits the positive impact we want to have on the world.

Cutting consumption down

Make no mistake, the effect on the climate of all the things we buy is huge. In this chapter, we'll break the issue of buying 'stuff' down into these five categories:

1. New clothing (potential saving - 2 tonnes).
2. New household furnishings (potential saving - 0.8 tonnes).
3. New gadgets and electronics (potential saving - 0.6 tonnes).
4. New books/magazines (potential saving - 0.4 tonnes).
5. New household appliances (potential saving - 0.2 tonnes).

The Ecological Footprint calculator tells us there's a potential prize of 4 tonnes of carbon footprint savings if you go from spending on the above regularly, to rarely. This would reduce the average UK person's CO2 footprint from 8.46 to 4.46 tonnes – down almost 50 percent!

By buying less stuff, you will cut down your personal footprint by a massive chunk (don't forget to go to footprintcalculator.org to measure your success). That said, it's a daunting prospect to change buying habits so drastically in one go. Lean/Agile teaches us to break issues down into manageable pieces and to prioritise them to make change palatable. So, pick from the suggestions and categories below, one at a time, depending on your lifestyle.

1. New clothing

Personally, I don't think there's anything wrong with wanting to look nice in the clothes we wear. But when it becomes a primary concern in our lives, it's probably not doing us any favours. If the CC22 has us buying new clothes all the time because the media preys on our insecurities, it could mean we're forsaking other activities that would bring us greater happiness. Going on personal experience, feeling like you need 'the perfect outfit for everything you do' is more of a stress and a worry than something that brings you joy.

I've gone from someone who was often in the shops obsessing about how clothes looked on me, to being far more laidback about buying new outfits. In terms of how it's affected my wellbeing, it's definitely a positive feeling to be free of that stress. I wouldn't want to go back to where I used to be.

Maybe that's not you, though. Maybe you'll never be happy in jeans and a t-shirt as you're someone who always wants to look well turned out. But, at the same time, you grapple with the environmental ramifications of buying new clothes. Perhaps you even got a bit of a shock to see that your footprint for buying new clothes could be as much as 2 tonnes a year. And, maybe buying clothes gives you ethical concerns due to the well-documented exploitation of workers in the producer countries. If so, then the solutions cater for you as well, because they're not about stopping doing what you enjoy, just about doing it thoughtfully, in a way that helps make the world a better place for us all.

The solutions

Host clothes swapping parties

You may well have swapped clothes with friends in the past, and it can make for an entertaining get-together – doing an at-home fashion show over a couple of glasses of wine. If you've never tried it, then it's definitely worth considering a party as a way to rejuvenate your wardrobe without hitting the shops.

Have swap parties with friends or even host something through work; it could be a great way to mix business with pleasure. Swap parties help build social and work relationships, and might even offer recognition for your innovative organisational and social skills! And, of course, it's a massive plus for the climate issue. This type of event is perfect for sparking conversations about how we can all play our role in avoiding catastrophic climate change. And, by swapping clothes, we're not adding to our footprints – the clothes have already been made, distributed, and sold!

Rejuvenate wardrobe faves

Most of us will have clothes in the cupboard that we can't bring ourselves to get rid of. They may have passed their prime, or we've simply worn them too many times because we love them so much. Instead of using up wardrobe space with forgotten treasures, you can get creative and spruce up much-loved clobber. If you don't mind pulling a needle and thread, and are looking for a positive pastime, then you could do it yourself. Or, you could use a (preferably local) business to breathe new life into them for you.

There are lots of creative people out there who would love to funk up tired clothes and make them *bespoke*, *interesting*, and *original*.

Although it's not mainstream right now, there will be someone who offers a rejuvenation and alteration service near you, for sure. If you're in the UK, check out clothes-doctor.com for an online service. You can also think about printing a design on plain clothing that you're bored of. Instead of trawling the shops, you could spend your time looking at artwork you enjoy to find a funky design. Maybe even you, or someone you know, is good at art and can create a custom design you'll be proud to wear.

Poppin' tags

Excuse the second reference to the Thrift Shop song, but I just love the sentiment in the lyrics that it's fun and original (ironically) to buy second-hand clothing.

There's stuff out there to fit any budget. From high street charity shops to vintage high street and online stores like rokit.co.uk, myvintage.uk, and rustyzipper.com; plus americanvintageusa.com in the US. As well as vintage fashion apps like Depop, and more commonly-known ones like eBay or Gumtree, there are also luxury pre-loved designer clothing websites like rebelle.com, or designer fashion apps like Vestiaire Collective. And auctions or donations of celebrities' pre-worn glad rags where the money even goes to a charitable cause, like at fashionforchange.boutique, can be enjoyed. Even Selfridges is trialling a re-sale initiative called 'Resellfridges' where customers can buy pre-loved fashion and jewellery.

Of course, it works both ways. If you want to get rid of anything, you can make some money by reselling. If you don't want to sell your used clothes, then make sure they go to charity or get recycled to give yourself a glow rather than guilt-trip from chucking them in the bin.

Capsule wardrobe

To cut down on how much you buy, it might help you to assemble a 'capsule wardrobe'. According to good-old Wikipedia, the term was coined in the 1970s and refers to a collection of essentials that won't go out of fashion, so you can wear them for years; each key piece can be worn interchangeably with just a few add-ins. This idea is fab for practicality like saving on wardrobe space, and makes choosing an outfit each day a doddle, as well as saving on pennies and making a big impact on your carbon footprint.

I won't go into the details here of the rules for putting one together, as it's easier to check it out online. The rules are laid out quite simply on Wikipedia if you need a first port of call.

If you choose to implement this improvement suggestion, then don't forget to do it in tandem with some of the others if it involves buying *more* clothes or disposing of old ones. That way, you will ensure you decrease your carbon footprint to the max.

Cut down bit by bit

If you buy a lot of new clothes, it's important to realise that you don't need to change overnight. Instead, you can use the key Lean principle of continuous improvement to buy less, until you reach your CO2e footprint goal. In reality, this means assessing how much you buy now and initially reducing it by an amount that doesn't seem daunting to you, so that you can bypass your fear of change. Each time you reduce the amount of new clothing, take motivation from the positive impact you are having. For more guidance on this approach, see the Appendix 'Continuous Improvement'.

You are what you buy

We're all bound to buy new clothes at least some of the time. When you do, you can choose to be selective. Choose to buy any new clothes from eco-friendly companies. There are lots of them around now in response to the ecological crisis we're living through, and in response to existing consumer demand.

The positive thing about a crisis is that it spurns innovation. It's a time when human beings can show what they're capable of. Crises have the capacity to bring out the best in us so we can respond in a way that brings us back from the brink of catastrophe.

Recent studies have even suggested that changes in the environment could alter cognition by modifying our neuronal connections. "Stress is the key factor in boosting plasticity, and learning, in the nervous system. In other words, when the going gets tough, the tough get going... It's only when you have skin in the game that you'll really focus and learn."[18]

In response to the damage that clothing, specifically, is doing to the natural world, an entirely new type of production has been created that considers the complete lifecycle of clothes.

It's called **circular fashion**, and it incorporates principles of Lean/Agile because it:

1. Gets rid of waste from the entire product lifecycle.

2. Is led by customers' requirements for quality and sustainable clothing that is *also* fair for workers.

3. Uses pull production – a Lean term that means something only gets made once the customer orders it (to avoid unwanted products).

Dr Anna Brismar, created the term 'circular fashion' and the website circularfashion.com offers a description. It says that fashion should be designed with "longevity, resource efficiency, non-toxicity, biodegradability, recyclability and good ethics in mind... Thereafter, the products should be redesigned to give the material and components new life. Lastly, the material ... should be recycled ... If unfit for recycling, the biological material should instead be composted to become nutrients for plants and other living organisms in the ecosystem."[19]

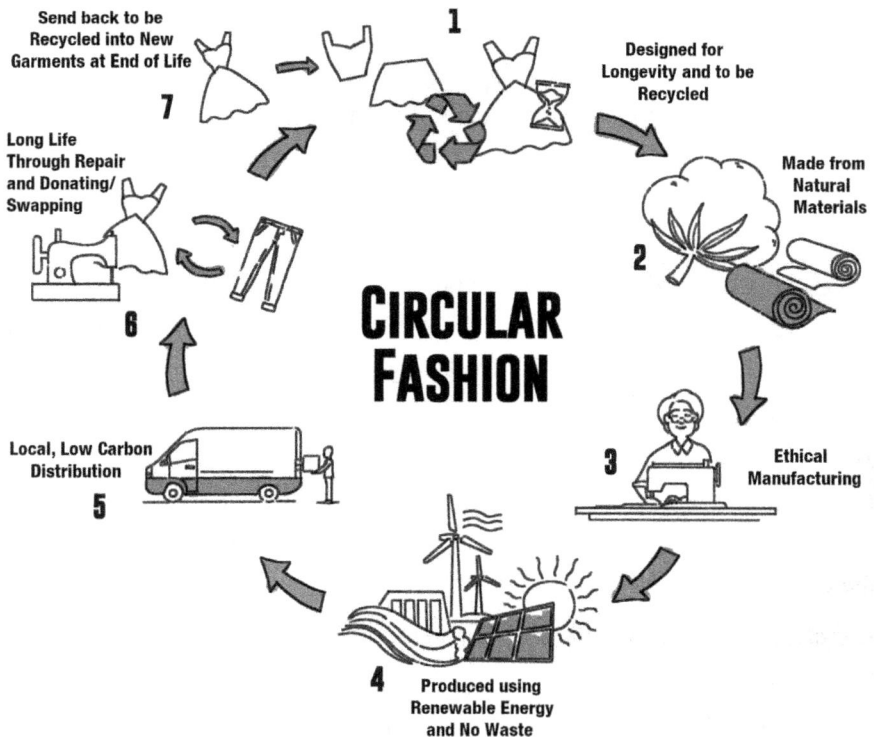

This type of production isn't just a pipe dream; regenerative businesses are already making it a reality. You can show your support by buying from them when you need to buy something new. I've recently bought from rapanuiclothing.com, for example. In turn, Stella McCartney is one well-known high-end brand that is incorporating these methods into their practices.[20]

You can also download the 'Good on you' app, which rates clothing companies in relation to their environmental impact, among other things.

Circular fashion isn't the only option for buying eco-friendly these days, but it is the complete solution that we should embrace to avoid unplanned side-effects from our eco 'improvements'. For example, making clothes out of plastic waste collected from the ocean seems like a fantastic initiative to reduce the detrimental effect of excess plastics on marine life. Big brands are getting on-board with this idea, including Adidas and Pharrell Williams' brand G-Star RAW. However, because these brands are not considering the *entire* lifecycle of the products, what happens to them 'in life' is overlooked. This may be making the problem even worse due to the number of microplastic fibres shed during washing, which make their way back into the oceans, and inside marine life.[21]

To avoid this unwelcome by-product, you could only buy things made from these materials (e.g., plastics) that don't need washing, such as accessories like bags and jewellery. As a rule of thumb, though, generally avoid buying synthetic fabrics as much as possible.

In summary, when you're buying new clothing, try to ensure the products follow the complete circular fashion model. Otherwise, it may be worth having a quick internet search on the company to find out whether the practices they use are causing negative ecological impacts. Bear them in mind before you hit buy.

2. Household furnishings

Consumerism sometimes has tragic consequences. It can sometimes make us forget who we are. In North London, a few years ago, a young man was stabbed and many crushed at the opening of an Ikea. There were big discounts on offer and, seemingly, the appeal of buying a soulless sofa at a cut-price rate was worth slashing another human being's body open for.

The over-packed, under-priced, warehouse-sized, spirit vacuums – that they call stores – can bring out the mob mentality in the best of us. Despite their lack of originality, style, or quality, many of the big chains appeal to us. We've created a world where convenience is king, and we're prepared to sacrifice our moral values for it. This is something I grapple with when I continue to buy things from Amazon, even having read articles highlighting staff working conditions, and having spoken to people first-hand who've had negative experiences working for them. I try to avoid it now, because it's not in line with my aspirational self, but I still use the site from time to time – despite my better judgement – because of the easy speed for things I feel I genuinely need.

I have found Ikea and other furniture chain stores a much easier habit to kick, and hopefully you will – once you consider alternatives and the damage that buying new furnishings for your home can cause. After all, you probably don't need to buy laminated coffee tables or any kind of furniture immediately; you definitely don't need to buy them new.

As most furnishings are made from wood, we have to make the connection in our minds that they used to be part of a forest. Ikea is purportedly the world's biggest consumer of wood, and according to a campaign against the company by SumOfUs, "Biodiverse forests in Ukraine are being illegally slashed to the ground to produce "sustainable" furniture for the Swedish giant."[22]

Clearing and burning forests releases huge amounts of carbon into the atmosphere and as Greenpeace puts it, "So much carbon is released that they contribute up to one-fifth of global man-made emissions, more than the world's entire transport sector."[23] So, not only is the notion of living in a world without forests and the creatures which inhabit them truly depressing, but the knock-on effect on the world's climate will be cataclysmic.

I visited my brother in-law's new home in 2019, and it was beautiful. The furniture looked brand new, but it was far from it. It was all beautifully restored pieces donated by their predecessors. The quality and character spoke volumes about their values, their history, and their passion for their living space. The feature wardrobe in the landing even had markings inside where ancestors had marked the heights of their kids growing up. It's a beautiful tribute to their families and the kind of self-expression that is full of meaningful gratification. With that in mind, take a look at the solutions below to save a hefty 0.8 tonnes of carbon from your footprint every year. You may also find that they improve your emotional wellbeing and give you a sense of purpose and achievement.

The solutions

Don't be an Ikea sheep, shop second-hand

There are lots of options for buying second-hand furniture. To name a few, there are:

- Apps such as Shpock.
- Websites like Gumtree.
- Antique furniture shops.
- Charity shops.
- Shops that specialise in restored furniture.

If you think about it, would you rather explore second-hand stores and sites for hidden treasures with a story? Or, traipse around soulless, brightly-lit stores for generic blocks? Personally, I think it looks stunning to have reclaimed wood shelving or a character antique table compared to new pieces that are poorer quality. In some cases, for the same or less money. Make this mindset switch; then you can rest in the knowledge that you're not responsible for desecrating living habitats when decking out your own home.

Restore/Upcycle

Maybe you have your own pieces of furniture that you know need love instead of being binned? Or perhaps you've bought (or have been donated) some lovely pieces that are full of potential? Either way, if you're looking for a creative pastime, then upcycling is a gratifying one, and there are a bucketload of videos on YouTube and other sites on how to do it.

Alternatively, take any furniture to a local business to restore. Give them your ideas and watch as a neglected piece of history becomes a beautiful piece of your life's fabric that can be enjoyed by your family now and in the future.

Restoring or upcycling your furniture is a great example of how we can reach the stretch goal of this book – to drive systemic transformation by implementing change collectively, as part of the Consumer-Led Movement.

The more people that replace new furniture with restoration, the more demand there will be for restoration services rather than new furniture stores. New industries and opportunities will emerge that centre around creativity and workmanship; instead of plundering natural resources to fill up stores with unloved furniture, depressed workers, and spoon-fed consumers.

Built to last

Inevitably, there will be some things that you'll want to buy new, either because you've done your best to find them second-hand with no luck, or because you want to be the first to own something where newness counts (like a mattress).

In these cases, there a couple of things you can do. Buy things made from responsible sources, and buy things that are built to last. The few minutes you spend looking online to find eco-friendly businesses are well worth it for the carbon saving. And, the extra money you may spend will be worth it for higher quality items that can be enjoyed for life.

3. Gadgets/electronics

If you're a real gadget geek, and get a big thrill from getting your hands on all the latest tech, that's no bad thing. But... there's a potential 0.6 tonne annual carbon saving if you go from buying new electronics all the time to rarely.

Perhaps you fly or drive rarely, and eat mainly plant-based food, and if you don't buy much other than electronics then you could find that you don't need to taper this one too much. As we talked about in chapter 1, technology can bring us so many advantages including being able to share information and harness global collective power like never before.

You may have sensed a 'however' coming... As with all types of consumerism, there is a difference between healthy pastimes and being sucked in by the system. If you're buying any new things regularly, it's always worth questioning if it's bringing real contentment, and what reasons lie behind it. For example, are we regularly upgrading our devices as soon as the newest version is out because we can make the most of the features they provide? Is it to use this knowledge to better our and others' situations? Or, has the latest slick advert made us feel a bit inadequate if we don't have the newest model out there?

If you think the answer is more likely to be the latter, and you want to reduce your unnecessary carbon footprint when it comes to electronics, then here's a few things you can do…

The solutions

Consolidate your devices

Once you feel less of a need to buy every new device, you can consider what you do buy, and how buying one device can give you the same benefits as several. Treehugger.com states that your phone is more than just a mobile, it's your "iPod, digital camera, electronic planner, e-reader, GPS device, calculator, and so, so much more. In fact, the more uses you find for your smartphone, the more environmentally-friendly it becomes."[24]

Buy second-hand/refurbished electronics

Even if you do choose to keep buying up-to-date tech because it's what you're into, you don't need to buy *new* new. Getting nearly new, refurbished devices can save you cash, and in the case of phones, SIM-only plans give you a better deal than getting a new device with a contract. By keeping virgin resources in the ground, and toxic waste out of the ground, both you and the planet are catered for.

Eco-friendly devices

It's worth knowing that not all devices are created equal. Search online for companies that consider their overall impact. Fairphone is a great example as it considers the entire circular economy, much like with the fashion industry, as previously mentioned. They make smartphones, and their website states, "It's got everything you'd expect from a great phone — and so much more. It improves the conditions of the people who make it and uses materials that are better for the planet. Because how it's made matters."[25]

4. Books/magazines

Just by regularly buying paper books and magazines, you create a whopping 0.4 tonne annual carbon footprint. Once you've digested that thought, it should hopefully be easy to make swaps to gain that carbon saving. I personally believe there's no need for any paper in our day and age. The swaps are all readily available and easy to do; everything is online, and it's just a case of breaking old habits. So:

- If you love reading paperback books, then consider buying second-hand (this one is printed on fully recycled paper if you're in the UK, and we use Print On Demand – similar to pull production – elsewhere).
- Read on your phone or tablet.
- If you feel you have to write on paper instead of capturing notes on a device, then buy recycled paper.
- If you can't bear catching up on news online then share any newspapers or magazines with friends, family, or neighbours to minimise your impact.

5. Household appliances

I can't imagine many people have a compulsion for frequently buying new household appliances. However, there's a potential 0.2 tonne carbon footprint saving to be made. Below are a few things to consider:

- Consider buying second-hand where possible.
- If you buy new, buy machines with top efficiency ratings.
- Buy appliances built to last.
- Take care of your appliances; get them serviced and fixed to avoid having to replace them.

The difference that coming together can make

Supporting businesses that provide products and services that focus on sustainability will help avert climate change and improve our environment. Especially if we do it collectively. Not only that, but shifting away from giving money to purely profit-driven businesses will also encourage system change away from maximising profit at any cost.

Kate Raworth, the author of "Doughnut Economics", believes that we have extraordinary opportunities to create a new economy. In her words, "Corporations that still pursue maximum rate of return for their shareholders suddenly look rather out of date next to social enterprises that are designed to generate multiple forms of value and share it with those throughout their networks."[26]

We can *choose* to support sustainable businesses like the ones mentioned; innovative start-ups that are using their heads to figure out a way to live in harmony with our world. These types of businesses tend not to use profit maximisation as their guiding principle. Supporting them will create demand for more businesses like them, and the profit-focused model of business will start to become obsolete.

Take the topics in this chapter as examples. Things like circular fashion and devices from companies like Fairphone. They consider the effect on employees and the environment – for the entire lifecycle – of producing, owning, and disposing of goods. Buy from them, and the demand will rise for these types of business; consumer expectations will shift so that they become the norm.

Instead of greed, which the current system perpetuates, in the new world, the shift in buying behaviours will focus our minds on planetary health and social wellbeing rather than profit. This will create more and more opportunities for a rewarding and innovative economy for us all. For example:

- If we cut down on buying new clothes and have them rejuvenated instead, there will be fewer jobs in large retail chains but more

- opportunities for individuals to learn in-demand design and textile skills.
- If we upcycle existing furniture, then even more people can make a career from providing a service that supports this sector, and we can leave our forests – the world's lungs – intact.
- Creating demand for refurbished phones will mean more people learn technical skills. Fewer people will work in retail mobile phone shops.

It is possible to drive meaningful, relatively rapid change with the collective power of our consumer choices. Both in terms of a combined reduction in our personal carbon footprint as well as driving wider business change. As Associate Professor Katherine Daniell says, "Cultural factors influence economic behaviour, political participation, social solidarity and value formation and evolution, which are closely linked to how and why public policies are developed in different ways in different countries." [27]

So a cultural shift created by us, in turn and in tandem with policy shifts, can move our entire system towards something that benefits us all.

In his fascinating TED talk, titled 'The Earth is full', Paul Gilding puts it like this, "I know the free-market fundamentalists will tell you that more growth, more stuff and 9 billion people going shopping is the best we can do. They're wrong. We can be more; we can be much more... We've built a powerful foundation of knowledge, science and technology, more than enough to build a society where 9 billion people can lead decent, meaningful and satisfying lives... This could be our finest hour."[28]

Key takeaways from this chapter

- Consumption is completely ingrained in us in modern society. This is enforced by constant advertisements that brainwash us into thinking that things = status and self-worth, which = happiness.
- In reality, buying things only has a very short-lived and hollow impact on making us feel better about ourselves; overall, it is more likely to make us feel worse. Often because consuming, and

thinking about how we can earn more to buy more, distracts us from doing the things that can bring us genuine satisfaction.

- By using our awareness of the CC22 to break free from it, we can make consumer choices that have a positive impact on our wellbeing. But not from buying what we're told to buy. Instead, we can choose to buy fewer things that don't add value to our lives. And what we do buy can represent the positive impact we want to have on the world, demonstrating to others our true values.

- Through buying new things rarely, instead of regularly, we can decrease our carbon footprint by as much as 4 tonnes per year (footprintcalculator.org).

- In the process, we can start to create a society where businesses need to adapt to our changing demands. This will mean more local, creative, and caring organisations emerge. And more opportunities to start one ourselves will appear.

4 | Transport

A while ago, I watched a YouTube video with comedian David Mitchell called 'climate change soapbox update'. In it, he poked fun at environmentalists who present sorting out the climate change issue as 'an opportunity'. The way he puts it sees environmentalists present sorting recycling as fun, whilst the Jeremy Clarksons of the world present driving a 4x4 to the North Pole – while drinking gin – as fun; and Clarkson wins.

Mitchell says, "His [way] is grotesquely irresponsible but like everything else that's grotesquely irresponsible – it's enormous fun. I want to see a global warming expert acknowledge that burning oil and the various machines we've invented that burn oil is brilliant, and it's a real pisser we can't do it anymore, but we can't. Because of facts."

It's totally tongue-in-cheek, and it did make me chuckle; I recommend looking it up for three minutes of light-hearted climate-related comedy (who knew there was such a thing!). As well as being amusing, the point that Mitchell makes is a valid one. There are some things we'll have to sacrifice for the sake of our future prosperity. Or at least they may seem like a sacrifice at first.

For me, transport has probably been the lifestyle category where the thought of making changes has been accompanied by a bit of a sting. Most likely because the convenience of getting around, and getting away in cars and planes, offers a sense of freedom. Doubtlessly, many of us yearn for freedom during the working week because such a big chunk of our lives is spent at work. Sure, some of us are lucky enough to have a job that provides some fulfilment but – for many people – work doesn't offer much more than a way to pay the bills. It's only natural we spend some of our time daydreaming about the next time we can drive somewhere and visit friends, or jet off somewhere far afield.

It takes a mindshift to see changing our flying and driving habits as a positive change to our lives. And, as with each category, it takes a deeper understanding of the problems our lifestyle is causing to make us willing and able to get there. Once we've digested the downsides of our current situation and have a glimpse of a better way, implementing the solutions becomes more of a pleasure than a chore.

The CC22 trap has us craving short-lived highs in fast cars, and we're left choking on the fumes and hankering for our next fix. Waking up from this – to a world where fulfilment comes from using the higher brain to create a brighter future – is a big win. Making conscious choices about what you spend your money on, to shape how we get there, is priceless.

So, if you're currently on team Mitchell, then take a look at the breakout box to help get yourself in the right frame of mind. But if you need no more motivation, you just need the practical steps, then skip it and head straight to the solutions.

Wrong today, gone tomorrow?

If we ever need evidence of how the CC22 is making us worse off, it's not hard to find. In chapter 1, we discovered that (thanks to our primal urge to be *numero uno*) economic growth has paved the way for maximising income. And without a second thought to the real cost to society in doing so. What we haven't explored is that the bigger and more damaging the business, the more ruthlessly this approach seems to be applied.

We're all familiar with the tobacco industry's cover-up once they knew the damage that smoking causes. With transport and climate change, it's well-documented that fossil fuel companies brainwash us with misleading advertising, lobby government to block low carbon policies, and fund climate change denier campaigns.

As Professor Robert Brulle is quoted as saying, "I think what we'll find is that the fossil fuel campaigns are going to dwarf what the tobacco industry did. It's an order of magnitude larger."[29]

Let's also not forget the scandal around diesel emissions tests being manipulated by car manufacturers – best typified by Volkswagen (but far wider) – so that it appeared they were far less harmful than they are.

Again, the lies we've been told in the name of profit claimed many lives. Volkswagen's excess emissions are projected to lead to 1,200 premature deaths in Europe.[30]

Air pollution from the transportation sector was thought to have caused 385,000 premature deaths in 2015.[31] "Concentrations from transportation emissions resulted in 7.8 million years of life lost and approximately $1 trillion (2015 US$) in health damages globally in 2015."[32]

What makes the situation even more nuts is that we now have the technologies that offer less harmful ways to fuel ourselves. For modes of transport, this ranges from electric cars and planes to hydrogen-powered boats. A better way is out there! If there wasn't a huge blocker stopping us then surely, we – governments, businesses, and individuals alike – would all be investing in a green economy. If our societal goals focussed on wellbeing over profit, there would be no question that tax subsidies should be put into sustainable transport instead of fossil fuels. Instead, direct fossil fuel subsidies are conservatively estimated at $20 billion per year in the US, and €55 billion in the EU.

"Fossil fuels account for 85 percent of all global subsidies... [Reducing these subsidies] would have lowered global carbon emissions by 28 percent and fossil fuel air pollution deaths by 46 percent, and increased government revenue by 3.8 percent of GDP."[33]

A report on the Global Subsidies Initiative website also says that a "subsidy swap" could make the clean energy revolution possible and save taxpayers' money. According to Richard Bridle, IISD Senior Policy Advisor, fossil fuel subsidies are often inefficient and costly to governments. They also undermine clean alternatives. Reallocating just 10 to 30 percent of fossil fuel subsidies would help pay for the much-needed transition to clean energy.[34]

As individuals, if the CC22 didn't have our priorities all messed up, then we wouldn't care about the status of our cars; regardless of how much fuel they guzzle. We could focus on what's best for our health and genuine contentment instead. And, we could readily move towards an economy that is fairer and more prosperous for all. (See chapter 9 for examples of how a sustainable economy could work.)

So… how do we get from A to B?

Our journeys determine our destination

Transport is the biggest driver of the average person's carbon footprint and therefore our most significant personal contribution to climate change. The main culprits are flying and driving. So, this chapter will primarily focus on solutions for us to travel differently so we can reach our 3-tonne emissions goal.

Based on the average person – assuming they don't carpool but do drive and fly – the CO2e footprint is 8.1 tonnes per year.[35]

We'll always explore the biggest emissions-impacting topics first which, in the case of transport, is flying. The average flyer spends 48 hours in the air according to HSBC research[36], which accounts for a whopping 5.3 tonnes of CO2e (source: footprintcalculator.org).[37]

Aviation accounts for 3% of global emissions, but on an individual level, regular flying is the worst thing you can do to contribute to climate change by a long way. I think we have to take the stance that although not everyone flies, if we support the growth of the airline industry by taking flights regularly, then we're assuming that more and more people will follow suit. And emissions from aircraft are predicted to triple by 2050 if demand continues to grow.[38]

For this reason, we need to consider what will happen if we promote more people making the same consumer choices as us. This is probably the right stance to take – morally speaking – as well, if you want to get the guilt fairy off your back.

Flying high

The HSBC research reveals that air travellers quoted multiple benefits to flying – including understanding the world better and being more tolerant and patient – as well as the personal benefits of feeling more confident and independent. I can totally relate to this having done some travelling as a young thing. As well as offering incredible experiences and delivering some of the most fun I've ever had, it opened my eyes to other cultures, and it gave me a feeling of a deeper kinship with people from all over the world. I think this is pretty invaluable to wellbeing on a personal level and for society. So I'm not going to be hypocritical and tell people they shouldn't ever travel and see the world.

I'm not even going to suggest to anyone that they shouldn't ever fly. What I am suggesting is that, within the next five years, we're aiming to reduce our flying time to fit within the 3-tonne goal. And the sooner you do it, the sooner you'll reduce your impact on the climate.

If you're a frequent flyer, the thought of dramatically cutting down might be terrifying. Even if you don't fly all that much, it's unlikely to be something you're keen to cut down on. And that's OK. It will be a process. Wherever you are on the scale, the guidance that follows should get you started on the right path.

When measuring your success, remember to use the guidance in chapter 2 as you implement any suggested improvements.

Flying for pleasure, don't waste time off

As I alluded to earlier, our time away from work is sacred. So, bearing in mind that we're aiming to increase our wellbeing as well as save the world, we should ask ourselves what we really want to get out of having time off. What is it about going on holiday that is going to add value to your life, and what bits might be chipping away at the value?

Most people are looking for things like relaxation and/or adventure from a holiday. Also, things like strengthening relationships, a bit of romance perhaps, or family time. Time to do the things you enjoy the most. Although

I haven't taken a survey, I believe with a high degree of certainty that virtually no one would proclaim that waiting around in airports and on runways is something they enjoy. We only put ourselves through it for the destination. Surely, then, it's worth considering some other options? Options that give us the biggest bang for our relaxation, adventure, or romance buck.

If you just want to relax, then do you need to go anywhere? Maybe chilling at home for a few days would invigorate you the most? This is an example of where a positive might come out of the Covid-19 lockdown. I know for me, personally, that taking time off just to 'be' was worthwhile. It was eye-opening to realise that.

It's easy to think that we're missing out if we don't jet off somewhere as soon as we have a few days off work. It's hard *not to* when advertising feeds us messages of how we're not living life to the full if we don't. In reality, *wherever* you are the world, there will be opportunities that offer all the excitement, comfort, and bonding you need without taking a flight. If, after some thought, you want to do away with the stress of travelling long distances, imagine the extra luxury you'll be able to afford and the time you can spend (having fun) that 'nearby' offers. Set yourself a little challenge to find the most unusual adventure, resort, or place of interest within a couple of hours' radius of your home. From pampering packages to tracking trails; I bet you'll be amazed at what you find.

Don't underestimate the benefit to your health of spending time in nature rather than on a plane. Aeroplanes are an incredible feat of human engineering that boggles my brain every time I see one. But being on one is probably about the furthest from nature I ever feel, and I don't just mean in the literal sense of being 35,000 feet from the nearest tree.

Spending time in pressurised, air-conditioned, flying tin cans is not linked to health. Spending time in nature, however, is. Recent studies have begun to show that being surrounded by trees and nature has a calming effect on our brains, and many other health benefits. Science has "begun to quantify what once seemed divine and mysterious. These measurements of everything from stress hormones, to heart rate, to brain waves, to protein markers, indicate that when we spend time in green space, there is something profound going on."[39]

Search for eco holidays in your neck of the woods to minimise your footprint even further, and get inspiration for the type of trip that will bring you closer to nature.

Weather the storm?

Being from the UK, I can't in all conscience overlook one of the key reasons why we want to go somewhere different on our hols (or vacays if you're in the US).

The weather!

If you long for a sunny beach trip or snowy mountains, and it means a lot to your wellbeing to have a holiday like that, then don't despair. You don't need to be in the Caribbean to sit on a beach or fly across the world for snow. There will probably be alternatives nearer to home that don't require flying.

For those of us who like to book trips away without using a travel broker, sites like seat61.com tell you how you can travel to specific destinations without flying. Moreover, it tells you about the adventure of the alternatives. How enjoyable the train journey through the Alps is, for example, with desirable stop-off locations en-route if you want to break up the journey and visit more places.

If you prefer a travel company to do the work for you, there are plenty of sustainable tourism companies who offer flight-free travel options within their packages. As Greentraveller.co.uk promises, "Of course, it is often quicker to fly but if you take the train, the journey becomes part of the holiday and can be a great adventure. Less Carbon, More Fun!"

Have a look on tourismdeclares.com to see which 93 travel companies have declared a climate emergency. This list shows some of the travel companies you could use, all in one place.

The sky is the limit

If you're desperate to go further afield and a flight is the only feasible way, then think about how you can fit a long-haul flight into your 3-tonne goal. It might be doable if you spread it out and avoid flying for a year or two afterwards. The point is to make conscious choices instead of the ones the CC22 encourages. Considering the value to you of taking a trip, and our goals of wellbeing and fighting climate change, ask yourself: do you really *want* and *need* to take a flight?

Personally, for the time being, I've factored in one return flight to Cyprus per year into my 3-tonne limit. This is because I have family there that I want to visit once a year and, at the moment, there is no other feasible way to get there. When this changes (which should be soon as a new ferry route is scheduled to open from Athens to Cyprus), I can reduce my flying time. My plan is to cut out flying completely if it becomes feasible to drive an electric car or get the train and take ferries for the entire journey. That said, I wouldn't rule out an occasional flight, for an important reason, as long as I factor it into my emissions limit.

Continuously improving

It took me a while to get my head around not flying on a whim. Initially, it feels like you're giving up holidays or the option to travel far afield. For me, this has included grappling with not visiting my sister and her family in the Caribbean where they have been living (I went once, a few years ago).

Some of the solutions in this book can make you feel a bit uncomfortable at first. If you're like most other humans, you feel uncomfortable because you're dealing with a quandary. You already knew that flying was bad for the environment, and having read this you may be feeling like you know it's time to do something about it. If you don't, you get to carry on flying, but you won't be living in line with your values of wanting to take care of the planet and our collective future on it. That's why having simple techniques that can make the transition from one mindset (and subsequent behaviour) to another is key. So, take a bit of time to digest and realise you don't need to immediately force changes you feel uncomfortable with. If you don't already feel you want to switch things up, keep reading about the effects of climate change, and explore the adventures you can have that don't involve flying until you do.

I can honestly now say that I'm excited for any future holidays I take. Partly because of the unusual destinations I'll discover that would have been sacrificed in favour of ones on a flight path. And also, the promise of adventure from travelling by means other than flying. The responsibletravel.com travel guide puts it like this, "…as soon as you realise that responsible travel is not about limiting ways in which you can explore the world, but actually about opening up layer after layer of our planet's potential, the excitement kicks in." [40]

Go to the Appendix section on 'Continuous Improvement' to find more guidance on using this approach. The 'Motivation' section will give you extra tips on how to take the sting out of not flying.

Flying for business

I've only done this once or twice, so I can't completely relate to all the reasons behind business travel. That said, I can relate to holding virtual work meetings instead of face-to-face ones. Over the last few years of my career, colleagues and stakeholders have been located in various towns and countries, and travelling is normally not feasible. What I've observed is that it's also often not necessary!

I'm writing this specific part of the book during lockdown due to Covid-19, so I don't doubt that many more of us have now experienced virtual meetings instead of travelling to meet people in person. Although no one would have wished things to change under these circumstances, I only hope that one positive to come out of the pandemic is that people are more comfortable with the idea of ditching unnecessary travel.

I hope that we're also more comfortable with utilising technology to its full advantage. There are multiple video conferencing software platforms available today. Two that are easy to use and secure are Microsoft Teams and Google Meet. Although 4G works well enough most of the time, 5G technology should soon mean that you can make use of them almost anywhere... without worrying about dropping your signal. Although the rollout of 5G will have some detrimental environmental impact, its capability is predicted to reduce CO_2 emissions by 1.5 gigatons by 2030 (over a third of the EU's emissions in 2012), even without taking a reduction in flights into account.[41]

In my experience, there isn't much that can't be achieved on a group call or video call instead of in person. And, in many cases, the benefits of avoiding frequent flying far outweigh the perceived downsides to calls or video-conferences. This applies to improving our health and work-life balance, as well as environmental advantages. Perhaps you are (or were) frequently flying for a company you work for, and you're concerned that you'll be expected to go back to doing it. If so, use this situation to initiate a conversation about how flying less frequently, and using technology instead, will benefit everyone going forward.

In his book "There is no Planet B", Mike Berners-Lee suggests that video conferencing does not reduce the number of flights we take. He says this is because of the 'rebound' effect, and gives the example that meeting someone on a video call can initiate further face-to-face meetings that would never have happened without it. He suggests a global 'carbon constraint' is needed to cap our overall energy use and resource consumption. Under this scenario, he says that video conferencing would become a vital tool to enable us to carry on as we are.

I also think a global carbon constraint is a good idea. However, while we're waiting for that to come about, having an individual carbon footprint limit will have the same effect on minimising our flying. Best of all, it's in our hands to do something about it right now.

Offsetting – can we pay the problem away?

Offsetting is worth a mention because it may seem like the perfect solution. You can carry on chugging out emissions because you pay to plant a few extra trees and – *voila* – you've got guilt-free travel. If you're not familiar with offsetting, or you're not sure why it isn't the answer to our climate change problem, these paragraphs aim to shed some light.

Firstly, what is offsetting? Offsetting means reducing greenhouse gas (GHG) emissions to compensate for emissions made elsewhere. Industry is often quoted as offsetting their emissions to meet 'net-zero' targets.

A big chunk of your carbon footprint is caused by deforestation to make way for agriculture. Or to mine for coal, or other minerals and resources, for the things you buy. To offset it, you could donate money towards planting enough trees that, in theory, would reduce GHG emissions by the same amount as your carbon footprint. Sounds great on the surface but the reality is it's not that simple.

Offsetting alone isn't the answer to the problem; for the same reasons that policy change alone isn't the answer. Offsetting initiatives don't tackle the root cause, and Lean/Agile teaches us that this is vital to problem-solving. This means these initiatives can never be the complete answer. They may treat some of the symptoms, but if the real sickness is our mindset and the system working against us going untreated, we're not going to get better.

I'm all for planting trees, but it isn't as simple as planting new trees for all the ones being cut down, and everything will be OK. A mature forest takes up to 100 years to develop, and anyone who's been into one will know that they are complex ecosystems of wildlife that have harmonised over centuries. And more than that, it is "…recovering the proportion of native species that are unique to the original forest which takes the longest time… up to 4,000 years."[42]

In a fascinating TED talk called 'How trees talk to each other', Suzanne Simard adds to this. She says, "A forest is much more than what you see… Underground there is… a world of infinite biological pathways that connect trees and allow them to communicate, and allow the forest to behave as if it's a single organism. It might remind you of a sort of intelligence."[43]

Listening to this, I couldn't help but be reminded of the movie Avatar. Suzanne explains how the older trees in a complex forest ecosystem nurture the younger ones, even of different species. And taking out too many of them through logging creates a tipping point that will bring down the whole system. Talking about our attempts at planting one or two species in their place, she says, "These simplified forests lack complexity, and they're really vulnerable to infections and bugs. And as climate changes, this is creating a perfect storm for extreme events."

Re-planting forests isn't the only way to offset carbon. The money you spend on offsetting could go towards renewable energy generation, for example. But whatever type of offset we make, we're in danger of thinking that it's a license to carry on producing carbon as we do now. This just isn't feasible on a large scale; offsetting will never keep pace with global emissions and resource use in a growing economy.

With all that said, offsetting *does* have its time and place. Although it clearly shouldn't be used as a get-out-of-jail-free card, there are situations where it's helpful. While we adapt our lives bit by bit to reduce our carbon footprints, there will be times when we can't manage to avoid carbon emissions. Flying may be a valid example of this. You may want to offset all your emissions at the same time as reducing them. Personally, once I reached the 3-tonne goal, it dawned on me to offset the rest.

Driving

In the UK, the average driver drove 7,600 miles in 2018, and the average petrol car's fuel efficiency was 36 miles per gallon (MPG). This makes the average person's (in the UK) footprint for driving (a petrol car) 2.8 tonnes per year according to footprintcalculator.org.

This figure is pretty scary. And if you drive, you might find the idea of cutting down even more alarming. I get it. The freedom, adventure, and sheer convenience that driving can bring are undeniable. Popping to see friends locally and visiting people and places a bit further afield in the comfort of a car is very appealing. Fortunately, done in the right way, (i.e., by following the suggestions in this section), it should be fairly easy to fit these things into your 3-tonne goal.

If you commute to work, then as I said in the 'flying for business' paragraphs, the lockdown may present an opportunity. Many of us are working from home instead of having to commute, and industry predictions suggest that more flexible working will be here to stay. Many businesses and people are feeling the benefits to wellbeing and productivity of working from home.

If you're not hopeful that your employer will be open to you working from home more often, it's worth *asking* the question and *stating* the benefits you feel it will bring to both parties. However, if this is a 'no go' then to reach the 3-tonne footprint goal within five years – depending on how far you travel and what you drive – you may need to factor in seeking employment closer to home.[44]

The things that affect how much carbon you're responsible for when driving (whether you own a car or are just using one), are how far you travel and how fuel-efficient the vehicle is (ignoring vehicle production and disposal for now). And every time you drive alone, all the carbon emissions are counted solely in your personal footprint. With this in mind, here are the improvements...

Don't stall - carpool

Perhaps you haven't considered carpooling as it might seem like a pain to do… but the upsides could make it very worthwhile. Benefits include:

- Every time you share a car journey, the carbon footprint of that journey is divided between the number of people in the car.
- The cost of petrol is shared.
- You get to meet and socialise with people.

If you normally do a school run, you can save yourself the headache of having to do it every day. One of my sisters has recently benefitted from this by linking up with a parent around the corner; now each of them only drives every other week, plus the kids (and probably you) get to make friends.

There are plenty of websites that make carpooling easier. BlaBlaCar.co.uk in the UK, for example, lets you choose the start point and destination and shows you rated drivers who are available and what the cost is. Ridesharing.com is one option if you're in the US or Canada where you can book your trip and pay online. Waze Carpool (owned by Google) is another option. It can also be used by the company where you work, for free, to help connect co-workers.[45]

Update your style

This suggestion is less of an obvious one. But it could reduce your carbon footprint by a lot and – if you're open-minded – be easier to implement.

An important Lean/Agile concept is that of 'flow'. It suggests that in order for things to run smoothly and cut out waste, anything that travels through a process should do so one at a time. Each item that travels through should also be suitably spaced out. This prevents bottlenecks and mistakes. Our intuition takes us down the wrong road because it leads us to batch things up, under the false pretence that it saves us time.

The flow principle is used in the production lines and processes of the most progressive businesses, and started out at Toyota. But there are countless examples of how this applies to our daily lives. Doing the washing up is a really simple one. We let it pile up thinking it's easier to do it in one go. But, in reality, this makes it difficult to manage and ends up meaning no room on the drainer and possibly more smashed glasses!

Another place we've got it wrong is the way we drive. We've convinced ourselves that the closer we are to the car in front, the quicker we get to where we're going. Not only is this incorrect, but it causes several issues, including:

- Accidents (many fatal).
- Personal stress.
- Traffic jams.
- More money on petrol.
- Increased emissions from road transport by as much as 15%.

I discovered this personally when I bought a car several years ago. At the time I bought it, the dashboard politely informed me that I was achieving 43 MPG. It was supposed to be an economical car getting much more than this, so I decided to test what would happen if I changed my driving habits. I began to leave more distance between me and the car in front. I left enough space so that I rarely had to brake because I was far enough back that just lifting my foot off the accelerator meant I didn't have to. I also minimised accelerating when I could see that I needed to slow down or stop. Before a junction or traffic lights, for example. After implementing these changes, within a few weeks, my MPG shot up to 58 – an extra 15 MPG.

Using UK stats as an indication, by doing this, the average driver would save around £250 a year in petrol (and even more through preventing extra wear and tear on the car), and reduce road casualties.[46]

These small changes in driving habits can have a huge impact on our immediate health, and reduce our impact on climate change. Each year, there are 108 million tonnes of greenhouse emissions from road transport in the UK. That means this driving change could save over 16.5 million tonnes of CO2 in the UK alone.

Walk that walk, peddle that bike

Back to the more obvious suggestions. You've no doubt heard them before, but they're worth pointing out because we can get into lazy habits of jumping in the car for the smallest of journeys without thinking. If doing this is taking you over your 3-tonne goal, and/or you just realise it's unnecessary, reap the health benefits of going for a walk or cycle instead.

For more tips that can help you put these suggestions into practice, go to the Appendix section on 'Motivation'.

Public transport

According to footprintcalculator.org, travelling the same distance by bus compared to driving saves you 0.4 tonnes of CO2e, and 1 tonne if travelling by train. Therefore, as a rule of thumb, public transport is the more environmentally friendly bet. It's worth bearing in mind that this may not ring true for you, personally. If you drive an electric car that is charged through renewable energy, for example, this will emit zero emissions on a given journey which fares far better than diesel trains or ones not powered by renewables.

Should we all buy electric cars?

Studies show that electric cars are better in terms of emissions than conventional diesel[47] and petrol[48] cars – even when not charged from renewables but *especially* when they are. This takes into account the carbon impact from the entire lifecycle of the car, including production and battery manufacture.

Driving an electric car run on renewables is not difficult in today's world, and the difference it makes to our footprint and the air quality around us is immense. We can take this even further by using 'systems thinking', which is just a fancy way of saying – solving problems by thinking about how things work *together*.

In this case, you could combine using an electric car with producing renewable energy at home. Mitsubishi have launched the 'Dendo Drive House' (DDH) system in Japan and Europe. This allows you to store the energy produced by your own solar panels in your car battery. This means you can "Reduce fuel costs by using solar panels to generate power during the day for charging EV/PHEV and domestic storage batteries, while at night, they can reduce power costs by using a bi-directional charger to supply power from their EV/PHEV to the home."[49]

In the UK, OVO energy and Nissan are trialling vehicle to grid (V2G) technology that enables you to store energy in the battery of your Nissan Leaf and send energy that is surplus to requirements back to the grid via an app.[50]

In the near future, thanks to 5G technology allowing local smart energy grids (more on this in chapter 6: 'Energy use at home'), you'll be able to share the excess with your local community.

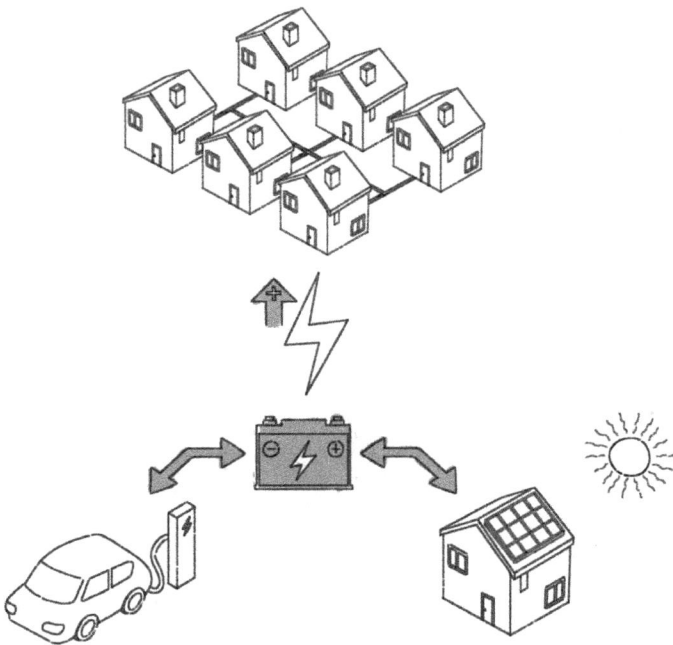

This technology and infrastructure is not a thing of the distant future, it's already here! It's only a matter of time before all this technology is integrated to maximise carbon reduction and give the power – quite literally – back to the people. The more of us that support the electric vehicle (EV) revolution, the quicker we'll see complete solutions. With emerging technology backed by consumer demand, there's nothing to stop the development of 'off-grid' communities. By combining what's already out there, it's very easy to imagine towns and cities where renewable energy is generated at home. It's then stored in home and EV batteries, to be used when needed, and any excess can be shared between individuals' homes and vehicles, and with the wider community.

Consumer demand for EVs is already huge. Be under no illusion, EVs are where things are headed. Electric car pioneer Tesla has shown that EVs are now big business. "The Tesla stock is now valued at more than $US380 billion by market cap, and is worth more than ExxonMobil, Shell and BP combined, as well as being the most valuable carmaker in the world."[51]

As much as fossil fuel companies hate to face it, fossil fuel-powered cars will soon be a thing of the past, so don't get saddled with one. If you can – make your next car electric.

Drive on?

Even with an electric car, endless driving without any thought about energy consumption isn't feasible in the long run. It's better to change your car to electric but with the view of cutting down unnecessary use (e.g., employing some of the other suggestions in this chapter alongside it). If you drive more efficiently and carpool or cut down where you can, you'll drastically decrease your energy use and make your car last longer.

You can also help shift our culture towards sustainability by considering the electric car that you buy. Think about buying second-hand instead of new. Also, consider whether the company that produces the vehicle takes into account the impact on the ecosystems and communities it touches during production and distribution. There is a helpful guide on the Ethical Consumer website, which is a great place to start.[52] Buy from these companies because they have goals outside simple profit maximisation, like minimising environmental impact and fairer working conditions. By supporting organisations that incorporate such goals, we're encouraging other businesses to follow suit.

Can we consume our way to sustainable flying and driving?

Flying – should we leave our future up in the air?

One option that's coming to fruition is electric planes. No one is exactly sure how the airline industry will be affected post-Covid-19, but easyJet did have plans to start testing electric planes for short-haul flights in 2023.

Another option is biofuel, which is made from plants. Second generation sustainable aviation fuel (SAF) means that the biofuel produced does not compete with land needed for food supplies or which is rich in biodiversity. The International Energy Agency (IEA) believes that blending these with traditional fuel will be essential for the aviation industry to meet its targets of reducing carbon emissions. The target is to reduce emissions to 50% below 2005 levels by 2050.

One other means of producing cleaner aviation is being trialled at Rotterdam airport. It works by capturing CO_2 from the air and mixing it with hydrogen at a plant, powered by solar panels. Dutch airline Transavia plans to be the first customer of this fuel, but it's currently too expensive to be produced at the scale required, and needs further development.

All these options sound great, but the speed of development of alternative fuels depends largely on social pressure. Cutting down our flying to fit within the 3-tonne goal is an effective way to demonstrate two things. One is that we won't continue flying thoughtlessly if we have to sacrifice our future for it; and that airlines must therefore focus on developing alternative fuels if they want to survive. It also indicates that we no longer support measuring societal success by long-term economic growth. And all industries must transition to something that guarantees a prosperous future without it. See chapter 9 for more on growth alternatives.

Just plane stupid?

One of the many practical reasons we can't continue to grow the airline industry – despite any efficiency gains or offsetting measures we may use – is that the amount of energy needed to get a passenger plane from a to b is immense. Regardless of how it's produced.

For example, if we use renewable energy and biofuels, there are emissions and resources associated with producing them. These will move us above the threshold the planet can take for a stable climate if we continue to grow the industry as planned.

Pre Covid-19, emissions from the airline industry were set to triple by 2050 "because of demand from people in developing economies to enjoy the same benefits of flying as those in rich countries."[53] Let's assume that after the lockdown ends completely, aviation industry growth will return to historical and predicted levels within the next few years. This is going to mean that any efficiencies gained[54] from more sustainable aircraft will be far outstripped by the rate of growth. Meeting the industry's CO2 goals will therefore require an increase in carbon offsetting. (We've already explored why offsetting is not the answer to our problems.)

Inclusive flying

If we reduce our flying in developed nations, we will be giving everyone the opportunity to fly. Occasionally. As part of a movement, we can begin to shift to a culture that prevents the airline industry from growing. And, we can support the development of alternative fuels so that we may continue to fly in the future.

If you're based in Europe, and pledging to go flight-free for a year sounds like your thing, then you can sign up to the 'flight-free' movement. The website in the UK is flightfree.co.uk. Maybe you're not ready for that pledge. If you want to incrementally reduce your footprint to fight climate change, and demonstrate to all major emissions-causing industries that you're taking part in collective action, sign up at consumerledmovement.com.

Driving – Are you miles ahead?

As mentioned, when using *systems thinking* to incorporate your EV into a smart grid, your car – and the system of charging around it – could offset more carbon than it creates in its lifetime. This is because less energy would be needed from fossil fuels to power other homes as well as your own. This could mean your EV is thought of as 'regenerative' or 'net positive' because it's saving more carbon emissions than it took to produce and use in its lifecycle.

So, as well as phasing out the production of fossil fuels more quickly, buying an EV could help speed up the much-needed culture shift towards 'regenerative'/'net positive' business models.

This awareness of how powerful our individual actions can be – when conducted collectively – is invaluable. But there's something else we must bear in mind. Earth has finite resources and planetary boundaries. And, in a growth economy with a capitalist ethos, the emphasis will be on producing more to maximise profit instead of what the ecosystems on our planet can cope with (before life becomes unstable).

To produce cars, this involves mining and using resources, *as well as* carbon emissions. This means there'll be an increasing impact on multiple ecosystems both from vehicle production and/or driving more. In a growing population, regardless of how much CO_2 is mitigated from renewably-powered transport and homes, this will ultimately become unsustainable.

We need to break out of the CC22 trap. To ensure our long-term survival, we should collectively shift culture not just towards sustainable technology, but also towards a system that doesn't focus on growth or self-interest. Having the 3-tonne goal is a good start to achieving this shift away from growth, because it makes us consider where we need to reduce our consumption, and think differently about having and wanting more.

Maybe, one day in the future, regenerative businesses will have advanced to such a degree that any amount of production and resource use *in all the ecosystems they touch* will only have a positive environmental impact. When that day comes, we can embrace growth again, if (and that's a big if) that's what we choose to do for the benefit of our wellbeing. In the meantime, we might have to put up with less shopping and more relaxing. Personally, I think I can handle that.

A final thought

If we're working from home, instead of travelling to the office, there is a carbon implication from powering homes *instead* of the office. Whether powering homes has a higher carbon impact than offices (which would offset the carbon benefit of the reduced commute) depends on the length of the commute. However, chapter 6 'Energy use at home' tells you how to minimise your carbon impact at home, which is also vital to achieving our 3-tonne CO2e goal. Following the suggestions in that chapter will ensure we maximise the benefit of cutting out unnecessary travel.

Key takeaways from this chapter

- Thinking differently about how we travel, and minimising how far we travel, might be one of the tougher areas to make changes in our lives. At least partly because of the sense of freedom and escape from the mundane it can offer.
- This is why the Lean/Agile problem-solving techniques are so crucial to our success in tackling climate change.
- Firstly, we can explore the root cause, and the damage it's doing, to give us the motivation to make refinements.
- In the case of transport that runs on fossil fuels, the desire for businesses to put profit first and for individuals to buy fuel-

guzzling cars, and go on exotic holidays, has led to us sacrificing the very air we breathe.

- We can consider what it is about travelling that adds value to our lives – like relaxing, visiting interesting places, strengthening relationships, and spending time in nature. Critically, we can ask ourselves whether we need to travel far afield to achieve that.
- We can also cut out what doesn't add value – which Lean teaches is waste – like long-haul flights and waiting around in airports.
- To reduce our footprint when driving, we can think about alternatives that can add value to our lives like taking a healthy walk or cycling instead, or meeting new people and reducing the amount of time behind a wheel through car-pooling.
- If we have the budget, we can think about buying an electric car and running it on renewable energy. And even take advantage of existing and soon-to-be technology to produce our own energy and use car batteries to store and share it. This progressive, systems thinking reduces our footprint and moves us 'off-grid'. It can also give us an eventual return on investment.
- That said, everyone buying an electric car alone won't lead us to climate safety. We need to think about how we can travel less, and therefore use less energy so that we can cut emissions to a manageable amount.
- On a macro level, this will mean moving towards an economy that doesn't rely on growth. And we'll explore this in more detail in chapter 9.

5 | Food

Hopefully, by now, you've been implementing some of the suggestions in this book to decrease your footprint. If so, you've probably begun to get to grips with the CC22, or how some of our less desirable traits that derive from our primitive brain have been compounded by capitalism – and this has created the mess we're in.

Our primitive or primal traits are not necessarily 'bad'. After all, they evolved for various reasons. But as some of those reasons aren't really applicable in the modern world, it's worth considering whether they're doing us more harm than good. We tend to dismiss our primal traits as 'human nature', but that's not really the case. Human nature – by definition – is what makes us human, and these characteristics are not that. What makes us human is higher, analytical thinking, but as the system doesn't encourage us to check in with our true selves, we have to mindfully take the initiative.

We've already talked about how the ethos of capitalism encourages our greedy, selfish side. And, in chapter 3, we talked about how incessant advertising preys on our insecurities because of the system's reliance on growth and consumption. But advertising also creates 'social norms', or widely accepted viewpoints and behaviours. This is thanks to its continual reinforcement of messages coming at us from every angle, AKA brainwashing. And, once we've made our minds up about something, we have another pesky trait that has a lot to answer for. It's called cognitive dissonance, and it plays a part in prolonging and exacerbating our problems.

Cognitive dissonance – we can be sure we're stubborn

Cognitive dissonance is the battle that goes on in our heads when we've formed a viewpoint and some new information comes along that contradicts it. This new information makes us feel uncomfortable because it undermines our current beliefs, and the way we live our lives based on them. So, initially, we'll scoff at the new information and reject it, no matter how rational it may be. After hearing it several times, it may begin to take hold, but it will make us feel very uncomfortable so we'll bury it or continually justify our original viewpoint against it. If we're open-minded and the new information seems valid then, eventually, we'll face it and change our minds and mould new values or beliefs.

Our stubbornness with our views can be helpful to us; otherwise we'd be constantly changing our minds at the drop of a hat which would make it near impossible to make any decisions. Left unchecked, though, our unwillingness to consider new possibilities in a world where we're constantly spoon-fed what we should eat in the name of profit is dangerous to our wellbeing, which of course includes the state of our natural home.

We are bombarded with adverts every day, telling us how healthy, delicious, and good it is to eat animal products. Advertising preys on our emotions with slogans like "Kill a cow, Start a fire, The magic begins", "You just can't beat this meat", and "Eat like a man" on meat adverts. Objectively speaking, of course, it's not true that eating meat makes someone 'manly'. Unless you're Bear Grylls, meat is picked up pre-slaughtered and pre-packaged from the supermarket shelf just like everything else we eat.

Years of media manipulation means that eating meat and animal products in large quantities, as part of a normal "healthy" lifestyle, is now part of our accepted worldview. Luckily, with scientific information emerging that indicates otherwise, this view is changing.

On mainstream news nowadays, we often hear that eating a lot of meat significantly impacts climate change and other aspects of the environment. As well as that, it has a detrimental effect on our health, and can cause suffering through the mistreatment of animals. If someone asked you whether you're ok with being part of the climate change problem, damaging your own health, and mistreating animals, I'm pretty confident the answer would be no!

But cognitive dissonance is a powerful blocker in our minds. So much so that even when we discover the burgers we're munching are marred with misery, rather than accept and apply that information to our behaviour, we'll do whatever we can to minimise, ignore, and dispute it. It's much easier to live with our heads in the sand than grapple with inconvenient changes; even when our current lifestyle choices go against our values!

Animal foods and health

Unfortunately, across the board, the scientific evidence on how diet relates to health is fractured and confusing. This means that all the information we hear or read can send our brains into a spin. When this happens, we end up sticking to our original viewpoint that has been ingrained in us, in part so we don't drive ourselves completely mad.

Having said that, I'll do my best to break down some key points in terms of animal foods and health objectively, to help us make those diets more digestible – figuratively, at least.

1. Many studies prove that animal foods contain nutrients and protein, which are beneficial for health.
2. Studies on diets high in animal foods have shown adverse effects on health. The World Health Organisation, for example, states that eating processed meat increases your risk of bowel cancer. Cancer

Research UK, likewise, has shown that 21% of bowel cancers and 3% off all cancers in the UK are caused by eating processed or red meat.[55]

3. The vast majority of studies and scientists agree that the public should include more plant-based foods, and less animal foods in their diet, to improve health and reduce the risk of disease.

4. All the nutrients and protein we need *can* be obtained from a *balanced* plant-based diet.

On this fourth point, the book "The China Study" details how early scientists have entrenched a cultural bias towards an extremely high protein diet, even though "There are virtually no nutrients in animal-based foods that are not better provided by plants."[56] The only exception is vitamin B12, which we no longer get from soil that's been depleted by the use of herbicides and pesticides (so someone on a plant-based diet may want to take a B12 supplement).

Food and health is such a complex subject and, for me, I find that reverting to common sense can help me make decisions when the evidence seems to be conflicting. In regards to our current high meat and dairy diet, it just seems logical to me that the modern diet must be far removed from what our ancestors would have eaten (before the meat and dairy industries became big business and were marketed to people constantly). In turn, I reflect on how it makes me feel when I eat a meal laden with meat and/or dairy, e.g., heavy, bloated, lethargic, and uncomfortable at best.

Personally, I had to give up red meat because my body couldn't handle it anymore. Just like you would cut out gluten if your body couldn't tolerate it; it wasn't worth the uncomfortable feeling it gave me physically, or in the case of meat – ethically as well.

Animal foods and cruelty

Although the link between eating animal foods and bad health feels blurred to many people, the link to animal cruelty is less so. We all know that, on the whole, animals bred for meat are not treated nicely and come to an untimely gruesome end. The sheer amount that we consume has made sure of that. Every one and a half years, more animals are slaughtered than the total number of humans who ever lived.[57]

It may be less obvious that animals bred for dairy are ill-treated but again, the amount of dairy products that are waved under our noses constantly has got us all hooked, and the demand can only be met through intensive farming. Any images supplied by marketers of happy cows enjoying life frolicking in fields, or tags like 'farm fresh' that imply animals live in disease-free conditions, are not even close to reality. The dairy cows that are able to roam outside still suffer from brutal systematic artificial insemination, and the immense emotional trauma of their new-born calf being taken away at birth, with the males often being shot. 95,000 dairy calves are shot in the UK every year.[58]

Eating eggs is no exception to cruelty either, regardless of whether you buy free-range. The male chicks do not grow fast enough to be kept alive for food, so they are killed when they are a few hours old.

A hard-hitting e-talk by Compassion in World Farming reveals that the secret weapon in food marketing is wilful ignorance by the consumer. This is another way of saying that cognitive dissonance may make us feel uncomfortable about eating animal products, but we ignore or suppress that feeling so we can carry on as we are. [59]

Cognitive dissonance and the CC22

Aristotle – the father of Western Philosophy – documented the conundrum of cognitive dissonance all the way back in c.340 BC. He talked about what happens when you see a sweet and want to eat it because it's tasty, but the rational part of you thinks about the consequences of eating the sweet and gaining weight. Adding appetite or desire to the equation overrules rational thinking.[60]

This observation has now been backed up by the neuroscience of what happens in the brain. Various parts of the brain are at play during cognitive dissonance. Parts of the prefrontal cortex – or the higher thinking, analytical brain – are trying to rationalise matters, but the medial frontal cortex (controlling survival instincts) and the amygdala (controlling emotions) can easily overpower things.[61]

Our rational brains are continually trying to bring the ethical and health implications of consuming animal products to the forefront of our thoughts. This is because these things matter *much more* than being told to think that greasy, ill-treated, hormone-laden chicken is 'too good to resist'.

But, as we have already covered, we know that the higher thinking mind is already suppressed due to the way capitalism operates. Then, add in the fact that we have this trait to reject or suppress new information (and that the primal brain overpowers the higher brain), and it's not at all surprising that we're still scoffing down sausages.

Eating animal foods and climate change

Of course, health and the potential mistreatment of animals are not the only areas of our wellbeing affected by eating animal products. It is now widely recognised and reported that our obsession with animal-based foods is a major factor aggravating climate change.

In all, direct and indirect emissions from livestock account for around 15 percent of all manmade greenhouse gas emissions globally. 7.1 gigatonnes of CO_2 equivalent per year. And a gigatonne is as big as it sounds. 1 gigatonne is 1,000,000,000 tonnes. [62] [63]

Making up 84 percent of this massive total is emissions from the methane released when animals digest food. And, the huge swathes of forest that are cleared to grow crops to feed the animals. As well as emissions from processing and producing their food.

(Manure storage and processing accounts for 10 percent. The remainder is due to processing and transporting animal products.)

Those who believe that changing our diets to be predominantly plant-based would have a similar impact on deforestation, should take into account that almost half the current global crop production goes to feeding livestock, and – on average – just 15 percent of these calories are then passed on to humans when we consume meat.[64]

Perhaps not surprisingly, considering the deforestation just mentioned, livestock agriculture is also one of the leading causal factors in the loss of biodiversity. And meat is considered one of the prime factors contributing to the current sixth mass extinction.[65]

In the UK, the impact of animal agriculture on our forests may be less obvious as they were cleared centuries ago. But the impact of animal agriculture on air pollution and our health is clear. A range of pollutant gasses and particulate matter is emitted into the atmosphere from fertilizer use, farm machinery, livestock waste, and livestock housing. But the main contributor is emissions of ammonia. And this has "effects on human health, increasing mortality and morbidity throughout the UK."[66] Also, scientists estimate that halving agricultural emissions globally could reduce the mortality attributed to the air pollution they cause by 250,000, and by 52,000 across Europe.[67]

A positive climate tipping point

If you're reading this book, you may well be someone who lies in-between the two extremes of a high meat diet and a plant-based one. If so, you may well be experiencing cognitive dissonance as you grapple every day with your true values and our social norms. In that sense, the climate issue may be the tipping point your willpower needs. Your rational health and ethical concerns may not have been enough in the past to overpower the primal parts of the brain. Especially when clever marketing and ease of access to animal products have manipulated it. Climate can be the part that tips the scales towards rational thinking.

"Becoming aware of the effect of cognitive dissonance on our decisions and understanding how we can overcome it can help us make better decisions and help us make positive behaviour changes rather than continue lying to ourselves."[68]

Now we know why we find it so hard to stop eating animal products, we can use our awareness to our advantage. Certainly for me, getting into the habit of considering how the CC22 affects my behaviour has helped create the mindshift I've needed. Now I find it much easier to make conscious choices about what I eat. These choices help prevent climate change and help me live closer in line with my values.

Solutions: the size of the footprint prize

Having a diet high in meat, which is normal in the developed world, produces around two tonnes more CO_2 than being vegan. That is a massive chunk of the average person's footprint. Go to footprintcalculator.org to take a fairly comprehensive dive into your personal diet and get a good idea which foods you eat are increasing your footprint.

The solutions that follow will suggest some of the ways you can reduce this whilst feeling that you're making positive, easily-achievable lifestyle changes.

Before we explore those, it's worth knowing that the intensity of emissions per kg produced is higher for dairy products like sheep and goat's milk than for meat products like pork and chicken. So, as individuals, we should be aware that products like Halloumi and Feta cheese are worse for our footprint than some meats. Not beef though; that's the worst on all counts.[69]

Plant the seeds of progress

We've established that the more we swap animal foods for plant foods, the more carbon we save from the atmosphere. The main solution that's going to help us reach our 3-tonne CO_2 goal is the bread and peanut butter of

Lean/Agile – Continuous (or iterative) Improvement. It can make the transition towards a plant-based diet a breeze.

I can speak from personal experience here, as I used to be a high meat eater. I'd eat meat twice a day normally, sometimes more. I was brought up in a household like many others, where eating meat was completely the norm. The issue of cognitive dissonance is also one that is very familiar to me. I remember – when I was young – scoffing at the notion of being precious about where meat comes from, or how it's produced, or the impact it has on animals' lives, despite considering myself to be an avid animal and nature lover. Lamb was one of my favourite dishes but – like pretty much all meat-eaters – I was completely detached. If I ever actually witnessed a lamb going to the slaughter, I would've been devastated.

That's why I can be so confident that *this* approach feels not only easy as you're doing it, but totally satisfying as well. I am not 100% plant-based yet, and I'm not sure if I will be, but I used to think that it would be virtually impossible for me to get to where I am today. As I'm writing this, I eat chicken about once a week (normally free-range from a local butcher), a (carefully sourced) egg or two a week, and perhaps a portion of dairy.

Even though my carbon footprint for my current diet is low, I'm continuing to decrease my intake, and I expect I will keep going until I'm vegan.

When applying continuous improvement, changes can snowball quickly. You may find that you naturally have more conversations and read things that keep you motivated. And you'll get ideas that make each step easier. New recipes, for example, or conversations about plant-based products and where to get them.

The footprint goal also gives you focus. You know the best thing you can do to remove emissions associated with diet is to move towards a plant-based one. So you can prioritise doing this above sourcing unpackaged or local food, for example. And, because it's one step at a time, you don't need to feel stressed thinking that you have to give up all the things you like, or that you need to learn to cook first, or figure out everything about nutrition and protein and get overwhelmed before you even start.

In fact, if you enjoy meat and dairy, I would recommend that you *don't* begin by thinking that you're going to give them up completely. Just think about cutting them down gradually to a point that's healthier for you and the natural world. You may find that, like me, the progression feels completely natural and you'll want to keep going… but you might not. It's a case of whatever you're happy with when it comes to that point.

For me, personally, cutting down on meat came first, probably in equal parts for climate, health, and ethical reasons. As my awareness and subsequent interest grew, I found it easier and easier to choose something vegetarian to eat, and when I felt the benefit physically – and with my overall wellbeing and peace of mind – it was a seamless transition to make those decisions more frequently. Going from eating red meat most days to rarely took about three years and then, a few months later, I was ready to give it up completely without it feeling like a sacrifice. That was over a year ago, and there's been a couple of occasions when I've felt tempted to have it. Reminding myself why I'm not eating red meat has helped a lot. On Christmas day 2019, I gave in to temptation and had a small piece which only served to remind me that it's not worth it.

With dairy and eggs, I'm amazed that I've got to a point where I rarely have them without feeling like I'm depriving myself. The way that I've achieved my dairy reduction is mainly by making swaps. Yoghurt for soy substitute; milk chocolate for dark; fake cheese, etc. Because it's been bit by bit and something that I've wanted to do for positive reasons, it has felt effortless. Now I can also focus on replacing the processed, packaged, and non-local foods with local wholefood without it seeming like a chore (more on this in chapter 8: 'Localisation', in the section on 'Local food').

How you decide to use the techniques to move towards a plant-based diet will, of course, depend on you. Maybe you love cooking and are looking forward to experimenting with some vegan dishes at home, or maybe you'll first swap out a weekly beef burger for a vegan one (these days there are some genuinely tasty ones). Either way, I hope you feel motivated enough to take some initial steps, or to keep going if you're already on the path.

(See the Appendix: 'Continuous Improvement' for more on using the approach.)

Origins

You can make a big difference when you're careful about where the animal foods that you buy come from. Some countries, particularly in Latin America, are even developing low-carbon livestock production. This will achieve emission reductions at scale, including reduced emission intensity, soil carbon and pasture restoration, and better recycling of by-products and waste.

"The world needs both consumers that are aware of their food choices and producers and companies that engage in low carbon development."[70]

With our food choices, we need to literally put our money where our mouth is. If you want to buy meat, but are on-board with buying less often, then this gives you greater means to be selective. If you haven't already, start to research and consider your local farms. And do a little research on their farming practices so you can back the ones that consider their impact on climate.

Food waste

In order to reduce our footprint enough to prevent catastrophic warming, we must tackle the topic of food waste. Seventy percent of people in the UK believe they have no food waste (wrap.org.uk), but sadly this is far from the case. In the UK, we throw away seven million tonnes a year (1.3 billion tonnes globally)[71], which costs the average UK household £840, or £70 a month... that's more than your average utility bill. Perhaps even more shockingly, five million tonnes out of the seven million tonnes of what we chuck in the UK is edible.[72]

Other wealthy countries fair even worse, but in all of them the percentage of household food waste sits between 30 and 40 percent and is the biggest cause of food waste from farm to plate. Apart from throwing money in the bin, this is a huge climate change-causing issue. Not just from the greenhouse gasses released as the food decomposes but because of the production, transportation, and storage of food that never gets eaten. Food waste has a huge carbon footprint of 4.4 billion tons of carbon equivalent, which is over 8 percent of total global emissions, and food thrown away at home is a massive chunk of that total.[73] [74]

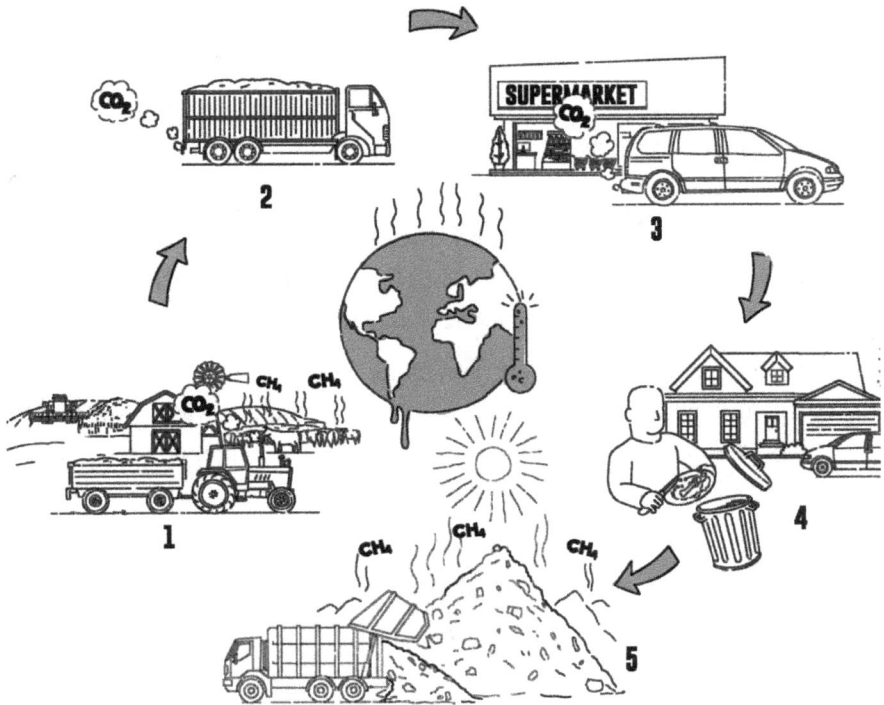

The footprint calculator we're using doesn't currently take food waste into account, so it assumes zero food waste. That means that you would need to add any wasted food to your carbon footprint reading at footptintcalculator.org, to give you a more realistic total. I've used a different carbon footprint calculator (https://mossy.earth/pages/carbon-calculator) to give us an idea that the average person will release about a tonne of CO_2 from food waste alone. Reassuringly, though, there are a tonne of solutions to cut this down that are an easy way to save money as well as carbon, and they satisfy more than just our appetite.

Solutions: Lean away from waste

In chapter 4, we explored the Lean concept of 'flow' which means preventing things becoming piled up and causing disruption and waste. Unfortunately, this is counter-intuitive to how we normally think, and as with many things, conventional thought is causing us problems. We are taught, or for some reason believe, that stocking up on things or doing things all in one go is beneficial and time-saving.

A great example of this is the way we food shop. We'll go to the supermarket as little as possible and stock up on things we think we might use. Then we fill up our cupboards and fridges with the stuff, which until we do use them is wasted time, money, and energy.

It's tempting to buy everything you think you need for the foreseeable future at once because it seems like the most efficient way to do it. The trouble is, it actually means that more mistakes are made, and food has to be paid for and stored without knowing if it will be consumed.

When you get home after an exhaustive shopping trip and unpack, you have a fridge and cupboards full of food. This feels great, until it starts rotting and you have to chuck it because what you thought you might eat in a few days' time, you never actually fancy. Or, your fridge is so full, you forget things are there. You then end up getting takeaways or extra food instead, which costs even more money.

Shop smart: buy little and often

This traditional way of shopping was nagging at me for ages as each time I threw away food, I got a little tug in my tummy from all the waste. After I began to take notice, it became so obvious that buying food in advance wasn't working; it felt pretty natural to do it less and less.

I find that as long as I have a few staples that I can use to knock up a meal, then everything else can be bought every couple of days as needed. Now I buy what I want when I want it, and because the fridge isn't rammed full, I can see what I've got. This means I can plan anything going out of date into the next couple of meals. It also means that I rarely throw anything edible away now.

Shopping little and often is also much quicker than traipsing around the supermarket for hours trying to imagine or plan what you'll be eating in six days' time, and there's next to no chance of forgetting anything or making mistakes. This means the whole experience is comparatively stress-free. On the occasions that I do bigger shops now, I only buy a small amount of fresh food and just buy the things I know will be eaten between shops. Also, I don't do a big shop while there are edible meals in the house. What is left is eaten first, and if there's a small number of ingredients needed to top up a meal, they're bought as required.

Storing isn't boring

Another thing that has a big impact on food waste at home is storing it correctly. I've become a dab hand at this now, and it's a huge help to stop scrapping leftovers. As soon as anything is opened, store it in the fridge (making sure your fridge is below 5° Centigrade) in an airtight container. I use the plastic containers that sometimes come with takeaways. These are perfect as they are reusable (as long as they have a number five in the triangle which all the ones I've seen do), airtight, and they are see-through. This means that as well as keeping food for longer, you can see what you have to use up. And, when you open the fridge, a quick glance tells you what ingredients you have to get creative with for dinner.

If you think you've got more than you need, then don't forget about the freezer. Go to lovefoodhatewaste.com and check out the storage A-Z which tells you exactly what you can freeze, and how to do it to keep the flavour.[75]

Get creative

If there are a few random things leftover in your fridge or cupboards to eat, get creative with them *before* you buy something new. Practically anything goes on top of toast or in a sandwich, and you might surprise yourself with a new favourite creation. I've recently got into making a kind of potato salad with leftovers which is basically a mish-mash of potatoes and whatever else I can find, with some olive oil and seasoning or some pesto. And they've all been really tasty!

Plan meals

Simply put, just have a little think ahead. Be conscious of the type of food you're throwing away and either don't buy it again, buy less of it, or plan it all into meals. Do you really need a bumper bag of potatoes? Or do half of them normally end up being chucked? Can you make use of them all next time? If not, stop. If you're opening a jar of pasta sauce but won't use it all that day, then plan *when* you next will use it to avoid forgetting about it until it's gone bad.

Veg boxes

A great way to eat more plant-based foods is to order a veggie box from somewhere that provides local produce with recipes that make use of it all. This has to be the most sustainable way towards a low carbon, ethical diet... as long as you plan the meals so that everything gets used.

Eating out

A simple tip to avoid wasting food when you eat out is to take your own Tupperware to make a doggie bag. The chef will love that you liked their dish so much you wanted to take leftovers home rather than waste them. Worst case scenario, you could actually feed them to your dog!

For a nice aftertaste

Go to lovefoodhatewaste.com for even more information and ideas on how to make sure edible food fulfils its purpose. It has recipes for what you can do with leftovers and other handy tips like an A-Z of storage advice and a portion calculator. Also, take a look at the Appendix section: 'Other Lean 'tricks' - 'How to' with less fuss and no muss' in the section titled 'The 5S System'. Here you'll find tips that will help make your personal food factory shipshape.

See the Recycling chapter for more, but for anything that isn't edible and the small amount of waste that happens because of life, don't let food end up in landfill – use food recycling or compost instead.

Tough choices today or tougher choices tomorrow?

Albert Einstein once said, "The significant problems we face cannot be solved by the same level of thinking which caused them."

Take farming cows, for example – a huge cause of greenhouse gas emissions. In a TV interview, author David Wallace-Wells spoke about a possible solution. It's been discovered that adding a small amount of seaweed to the cows' food could reduce methane production from the cows by up to 99 percent. He suggested that if this becomes a policy enforced by governments, the issue of cow farming in regards to climate change could be resolved.

I can see the appeal of this kind of solution; it basically means that everyone can carry on as they are, no scary change needed. Just pass the buck to governments and carry on chowing down on burgers. The trouble is that this kind of solution hasn't come from understanding the root cause of the problem. If we begin to address the root cause – the CC22 – it will give us the opportunity to use our higher thinking brains. This means the solutions will address the real problem and create lasting positive change that increases our wellbeing. But, with solutions that rely solely on policy change, the public will continue to eat beef and dairy in the amounts we do now, causing the following:

- Deforestation.
- Desertification.

- Overuse/mismanagement of resources needed elsewhere, such as water.
- Human health issues.
- Mistreatment of animals from mass production.

Therefore, we would completely miss the point.

I want to reiterate here that I advocate policy change. It has to happen, and if you're so inclined, then absolutely go out and protest. Be an activist and write to your MP, and anything else you can do to drive policy change. But, if you do all that without considering how you live your own life, then surely you don't appreciate how destructive we are all being to our planet? The CC22 root cause demonstrates that it's the capitalist system *and* the lesser traits of our psyche that have caused climate change. Political change alone will not resolve the climate issue. In order to affect the outcome of climate change, we must also look at ourselves.

As individuals, the continuous improvement approach will make changes seem easy. And it will also give the livestock industry a chance to adapt to consumer demand which will lessen unwanted effects such as job losses.

That said, for the amount of change required in our diets to prevent catastrophic climate change, people's lives will be impacted. In the livestock industry, businesses will have to adapt quickly to either:

- Be part of the sustainable livestock sector, and cater for the remaining animal foods demand.
- Move to plant agriculture to meet the growing demand.
- Change sector altogether.

In my opinion, this is where we do need the government to be the driver. We will need them to back the transition from meat to plant-based diets with subsidies and whatever training and support livestock sector businesses and workers need. We shouldn't underestimate the importance of doing this properly to stimulate change and avoid negative impacts.

There are people who argue that the loss of jobs in livestock agriculture, and subsequent impacts on the economy are reasons not to move away from animal agriculture.[76] These people need to consider the overall devastating effects on society if we don't. Firstly, that "climate change is one of the biggest threats to economic stability."[77] But also that things go way beyond job losses (if not managed effectively by government) in one sector. Climate change will negatively affect the quality of life for the entire human species.

Carrying on as we are will mean everyone on the planet will ultimately suffer, starting (as it already has) with the world's poorest and most vulnerable. The heatwaves, droughts, and floods – even in the UK – will be responsible for taking innumerable lives and causing extreme suffering. On *just one day* in the UK in 2019, which was a record-breaking 38.7° Centigrade, more than 200 additional people died according to the Office for National Statistics.[78]

And things are going to get much worse. In 50 years' time, "For the EU-27 Member States, the PESETA study projected almost 86,000 net extra deaths per year in 2071-2100, compared with the 1961-1990 EU-25 average, for a high emissions scenario (IPCC SRES A2) with a global mean temperature increase of 3°C."[79]

It might seem somewhat abstract to quote these statistics, so I think it's worth picturing yourself at 70 or 80, whether that be in 20 or 50 years' time, and imagine how unpleasant it will be to suffer at the hands of extreme temperatures, which will then be the norm. Imagine not just the suffering caused by the temperature, but also by the knock-on effects of scarcity of water and a health service under vastly more pressure than today (the Covid-19 situation has given us a little taste of what this feels like). This is the best-case scenario we're being given if we don't do something about it now.

It would be amazing if all the evidence about climate change is wrong. I want my son and younger loved ones to grow up in a world that isn't facing this civilisation-ending threat, where they get to enjoy the same natural wonders that I did.

But I cannot, and we should not, ignore what the mountain of evidence is telling us. To suggest that we shouldn't move away from eating meat regularly on the tiniest chance that the effects of climate change are being overstated, because there will be some job losses incurred, seems mind-boggling. If you sit in this camp, then be open to the idea that cognitive dissonance may be clouding your better judgement.

In regards to what we eat, I believe that tragically, we are collectively in danger of letting our stubbornness be the cause of immense global suffering. Because we're resisting the logical and more ethical things to do.

Key takeaways from this chapter

- Cognitive dissonance is our stubbornness to accept a viewpoint that doesn't match our existing one.
- This happens even though our existing view may be a result of cultural conditioning – like adverts reinforcing the message to buy animal products – rather than the values we hold dear when we're accessing our higher, human brains (like our health, the welfare of animals, and the effect of our actions on the environment).
- Emissions from livestock account for 15 percent of global GHGs. But understanding our obsession with animal products gives us the power to break out of our bad habits and unhelpful mindsets to think and act differently.
- This can help us cut down on animal products, bit by bit, and save as much as 2 tonnes of CO_2e per year.
- You can also save around a tonne of CO_2e by making sure the food you buy gets eaten rather than thrown away. Use Lean Thinking to buy food as you need it, and make good use of what you have.

- We can apply our understanding of cognitive dissonance to the future of the food industry as a whole, and try to be open-minded about the vision of where it needs to go.
- We can accept that although there will be challenges that we'll need to overcome, we must move away from animal agriculture and towards a diet that increases our health, respects the Earth's creatures, and keeps us within the limits of climate safety.

6 | Energy Use at Home

Easy, not sexy

Energy use at home probably isn't the sexiest of topics to cover, but it may be the easiest one for making huge carbon savings and steering the energy industry in the right direction. And, we can do all that with very little effort.

One reason we can make BIG carbon savings is that the majority of people's homes are still powered using dirty energy, AKA fossil fuels. And fossil fuels are the dominant cause of the globally catastrophic climate change that looms over us.

The research mentioned in chapter one showed that individual consumers are responsible for the biggest impact on greenhouse gas emissions (GHGs). As well as the things we buy – which have carbon footprints that we're indirectly responsible for – the energy we use for powering our lives has a direct effect on climate change. Specifically, a massive 20% of *all carbon impacts* are down to how we fuel our cars and homes.[80]

Being dirty isn't all bad

Fossil fuels are now seen as being 'dirty' energy. But, nonetheless, they are the helping hand that has been – very kindly – provided by the ultimate nurturing parent: Mother Nature.

The Earth has been busy cultivating enough coal, oil, and gas reserves, over the few million years preceding our arrival on this planet, to literally fuel our progress as a species. This has allowed us to develop technologies at incredible speeds since the industrial revolution.

About the era that followed industrialisation, scientist Geoffrey Ozin says, "Inventive ideas transformed into technologies that sought... to exploit the strength of steel and power of electricity to transform early modes of transportation, manufacturing, communications, and agriculture to modern versions..."[81]

But the damage that fossil fuel extraction and burning causes makes it seem clear that it was only ever meant to be a stepping stone. It's been a brief flutter in time since it was discovered and harnessed.

Yet, in that time, we've used it to develop technologies that mean we are *now* able to harness and use renewable energy *on a global scale*, which is far more sustainable. In that same brief moment in history, both the fossil fuel reserves, and the capacity of humanity to survive using them, is running out. It seems almost poetic that the planet has provided us with the amount of energy we've needed to propel ourselves to a higher quality of life, for the amount of time we've needed to do it.

But, we have reached the fork in the road. The scientific predictions are not a crystal ball, but they are crystal-clear. For every day, week, and year that we continue to use fossil fuels, we're pumping more harmful gasses into the atmosphere than the Earth's ecosystems can store, and this is worsening climate change. If we carry on using them, we'll have to rely on technology to suck out excess carbon from the atmosphere. But this technology neither exists nor is likely to work on the scale required.[82]

In other words, we'll be in a position where we're unlikely to avoid globally catastrophic warming, which will cause immense suffering to human life. This suffering comes from both the direct effects of extreme weather events and sea-level rise, and knock-on effects such as disruption of food production and water distribution, and war.

The future can be bright

The good news is that we don't have to rely on fossil fuels going forward. Many experts have proposed plans for a future without them. In fact, there are 42 peer-reviewed 100% (or near-100%) *global renewables* studies by academics and Non-Government Organisations (NGOs) alike.[83]

Mark Jacobson, from Stanford University, co-wrote a plan to get 139 countries of the world onto 80% renewables by 2030, and 100% by 2050. In his summary, he explains, "We develop roadmaps to transform the all-purpose energy infrastructures... to ones powered by wind, water, and sunlight (WWS). Converting may create 24.3 million more permanent, full-time jobs than jobs lost. It may avoid 4.6 million/year premature air-pollution deaths... Transitioning should also stabilize energy prices... increase access to energy by decentralizing power, and avoid 1.5°C global warming."[84]

Tragically, the influence that fossil fuel companies have on governments means that administrations are not putting policies in place to make these changes come about quickly enough to avert disaster. But our energy use at home is one clear area where we can use our collective power as consumers.

Our individual consumer choices to support sustainable energy will also impact how businesses operate. Businesses must adapt to consumer demand or they die. That means that our investment in the *right kinds* of energy businesses can change the face of energy production.

Businesses are already consumer-led, but we're not taking full advantage of that. We're letting advertisers and the media brainwash us into paying for things that harm our wellbeing. And cause climate change.

Scientists from ExxonMobil – one of the world's largest oil companies – created one of the first models to predict the devastating effects of man-made climate change. But instead of communicating this to the world, Exxon set about creating doubt in the public sphere using the same 'blueprint' that the tobacco industry did, pitting (climate) scientists against (bought or biased) scientists. And they ran advertising campaigns. A recent lawsuit instigated by the General Attorney of Minnesota claims that despite industry knowledge, Exxon (among others), "engaged in a public-relations campaign that was not only false, but also highly effective," which served to "deliberately [undermine] the science" of climate change."[85]

This is why it's so important to be a voice in our local and global community for making more conscious choices. As soon as we choose to do this collectively, the world will truly be consumer-led. Businesses and organisations will have to develop and adapt themselves to serve our true needs, because we demand it.

If you want to add more fuel to your fire before you get going on the improvements, then read the breakout box to learn more about how dirty fossil fuel companies really are. If you're already all fired up, then skip it and head straight to the suggestions.

Push policy

There was a televised climate change debate before the 2019 general election in the UK. The parties present made it clear that they supported moving to renewables and away from fossil fuels. Retrofitting houses for energy efficiency was both high on the agenda and budgeted for.

Unfortunately, the one party that got into power was the one that didn't bother to show up to that debate. Looking at their manifesto on climate, it's not surprising why they weren't there. The Conservative Party have said they will deliver with regards to achieving net-zero emissions by 2050 (although the other main parties pledged to do it sooner), but in regards to fossil fuels, they stated, "We believe that the North Sea oil and gas industry has a long future ahead and know the sector has a key role to play as we move to a Net Zero economy."[86]

You may have heard in the press recently that major fossil fuel companies have also pledged to achieve net-zero emissions by 2050. What they, and the Conservative party – along with many governments around the world – have failed to demonstrate is *how*.

On BP's statement for getting to net-zero emissions, a Guardian article noted, "There are no concrete details here about scaling down production of fossil fuels or scaling up renewables. This will raise concerns that the company thinks it can just plant trees or use other offsets to make up for ever-greater petrochemical production. This would not be enough to stabilise the climate."[87]

The UK was one of the first countries to declare a Climate Emergency in 2019. This acknowledgement of the size of the issue is great, but without policies to back it up, it's little more than hot air. According to WWF, action to back up the declared climate emergency includes *ending* support for fossil fuels which – as just mentioned – the current government in the UK does not intend to do.

Action also includes stopping pollution from our homes. In this case, as well, the current administration offers no comfort that they are taking either the declared Climate Emergency, or their net-zero emissions targets, seriously. What might measures look like? Using renewable sources, the installation of heat pumps and solar hot water, and ensuring all new homes are zero-carbon. Instead, the Conservatives support gas for heating and talk vaguely about creating "new kinds of homes that have low energy bills."[88]

We all know that gas is a fossil fuel, but (in the UK) it's also the fuel that's most used for both heating *and* electricity generation. Although coal is the dirtiest fossil fuel, natural gas is more prevalent now, and it accounts for a fifth of the world's total carbon emissions. It is not a clean alternative. And if it continues to be supported by the government, we cannot reasonably expect to meet the carbon targets we've set, let alone decarbonise sooner.[89]

"Policy development has begun for many of the components needed to reach net-zero GHG emissions... These policies must be strengthened and they must deliver action. A net-zero GHG target is not credible unless policy is ramped up significantly."[90]

I've talked about the UK a lot so far, partly because I'm from there but also because it's an interesting example. The UK is ahead of most places in terms of its political commitments to tackle climate change. Sadly, considering everything we've just explored, that's not very comforting. This is the reason why, in developed countries, we must work collectively to push through the changes needed.

Mind over money, matters

The thing stopping us from making changes to our consumer behaviour is the Capitalism Catch-22. The system has us distracted. We have busy lives involving lots of work and little play, so don't have the headspace to spend time considering the damage fossil fuels are doing. That's why when the big fossil fuel companies announce plans for net-zero emissions, we're in danger of thinking – *phew*, time to stop worrying about what they're up to. *They're on the case.*

But big fuel company bosses and shareholders are the biggest victims of the CC22. They have allowed the system to emphasise their greedy, selfish traits coming from their primal brain. And they've done this to the point of actively concealing and perverting the truth about the damage fossil fuels cause, in the name of profit.

"Many of these companies have also strenuously lobbied – directly or through influential industry trade associations – to block policies encouraging the needed transition to low-carbon energy."[91]

So, deep down, if you have a feeling that big energy companies taking the necessary action sounds too good to be true, it's because it is. At present, they should not be trusted to look after our future, because they're more concerned with looking after their present.

It's undeniable that if we work together, the rest of us (billions of people) have more clout. So we have to ask ourselves, are we going to let the mindset – caused by the system – seal our fate? Of course not! Let's explore the options for reducing our carbon footprint by managing our energy use at home.

How much carbon do we produce at home?

Someone in a medium-sized home[92] can reduce their carbon footprint by 2.1 tonnes per year (footprintcalculator.org). This depends on whether:

1. Their energy sources are renewable.
2. Their home is energy efficient.

In the best-case scenario, someone could go from 0 to 100% renewables and from a very inefficient home to one with an energy-efficient design. This would reduce the average footprint per person in the UK from 8.46 tonnes to 6.36 tonnes. That's 24.82%; just shy of a quarter! Once you've checked your carbon footprint at footprintcalculator.org, you'll be able to see what reducing your footprint by up to 2.1 tonnes means for you personally, to reach your 3-tonne goal.

As always, we'll prioritise things, so you can tackle the solutions that will make the biggest impact first.

- By using 100% renewable energy, you can reduce your footprint by 1.3 tonnes per year.
- By having an energy-efficient home, you can reduce it by 0.8 tonnes per year.

Now let's break down the improvements into manageable chunks to make the prospect of achieving this go from daunting, to easily doable.

Switch your supplier

I'm betting that a lot of people reading this are already using a renewable energy supplier. If you are, then you can skip to the next subheading titled 'Fifty shades of green', as is turns out that's pretty relevant in regards to renewable energy tariffs. In other words, there are still some things you may want to know. But, for those of you that aren't yet on a renewable energy tariff, this is the easiest win in the whole book. You'll knock 1.3 tonnes off of your carbon footprint, or more if you have a bigger-than-average home, and all in about 15 minutes!

All you need to do is *switch*. Go on a comparison site, answer a few quick questions about the house you live in, and sign up to the cheapest or (perhaps more importantly) best (criteria coming up) renewable supplier. Then the energy companies take care of the rest. At least that's how it is in the UK; chances are, whatever country you're reading this in will have something similar.

There's no doubt that making the switch and reducing your footprint by such a huge chunk will make you feel good about yourself. My last renewable electricity supplier informed me that the average member saves 1.31 tonnes of CO_2 per year, which is like planting 655 trees – a very mood-boosting thought. And every time you think about using electricity, you'll get a little buzz from remembering that you're not using fossil fuels to do it.

See the Appendix section 'Motivation', in the section 'Quick productivity tip', for help to make sure you get this done asap.

Fifty shades of green

If you have a centralised distribution system for electricity, or 'national grid', where you live, then the energy that comes to your home will be a mix of the sources that have supplied it. Some will likely be renewable, e.g., solar, wind and hydro, and some from fossil fuels.

I think most people can accept that there isn't a direct line from the renewable supplier who generates the electricity and the buyer's house. I've seen some people comment on social media that the electricity mix between renewable and non-renewable coming to your home via the grid is a reason not to bother using a renewable supplier. But I don't see their logic; probably because that argument doesn't use much.

If you're paying for renewable energy, then you're voting with your wallet for renewable energy and the investment, development, and use of it will continue to increase. This means the sources of electricity entering the grid will be ever more renewable until fossil fuels are no longer part of the mix.

Things will continue to transform and improve so long as we're heading towards the right goals. Perhaps we'll move towards decentralised energy as we continue to adapt to a way of life that mitigates climate change. Some places are making the most of this approach already, where locally-produced renewable electricity is used and shared in the community.

The current approach – of buying renewable energy from a central grid – might not be the end state, but don't shun it because it's not perfect. For the time being, it works because all types of renewable energy are a significant improvement on the use of fossil fuels.

There are some things about renewable energy companies that we should be mindful of. Because their success *will* affect how quickly and successfully we stop using dirty energy. When choosing a renewable energy supplier and tariff, there are some questions we should ask:

1. Does the supplier also sell **non**-renewable electricity?
2. Does any arm of the company produce fossil fuels?

If the answer to either of those is *yes*, then try to avoid that supplier. Some companies use a mix of renewable energy and fossil fuels for electricity generation (Gas is a little different as we'll explore next). Or, they offer renewable-only electricity tariffs but also produce fossil fuels. If that's the case, then there are other energy suppliers which only support renewables. We may be better off giving our money to these companies to make the transition to renewables as speedy as possible.

Suppliers with social goals

When choosing a supplier, you may also want to consider whether they use some or all of their profits to invest in further renewable energy generation, or tackling climate change.

Only a handful of renewable energy suppliers in the UK do this. If they don't, they're probably not the type of business to support going forward in the fight against climate change. We need things to happen as quickly as possible, and investment in tackling the issue will help speed things up. Also, we want to shape businesses' offerings and the way they operate for a long-term sustainable future. One crucial means to make that transition is by supporting organisations that have social and environmental goals as well as, or instead of, just profit. (See chapter 9 for more on this.)

Finding the answers to these questions may involve an extra few minutes of online research, but the companies that do invest in renewables or – have goals other than profit – tend to shout pretty loudly about it on their websites, so it's not hard to find out. In some cases, you may pay a bit more for these suppliers.

It's worth bearing in mind that it's not the world's poorest that are the main contributors to climate change. The people who can afford to keep their ample-sized homes toasty in winter and cool in summer are the main culprits. What I'm saying is that if the reality is that we *can afford to*, maybe it's OK to pay a little more to bring about the type of world that will serve the whole global community in the future. Not only that, but the more we use and continue to develop renewable energy sources, the cheaper it becomes. So, if it's not already cheaper than fossil fuels where you are, get behind it, and it soon will be!

Gas

In places where people use gas as part of the energy mix to fuel their homes, finding a renewable alternative is a bit more problematic. In the next section, on having an energy-efficient home, we'll explore some alternatives to gas-powered homes. For now, to gain an immediate carbon benefit, you could use a supplier that offsets its gas or uses a proportion of more sustainable/renewable sources, AKA green gas.

Biomass, for example, is the plant or animal material used to produce green gas. Sources include food or farm waste. But those specific sources probably won't work as long-term solutions. We need to dramatically reduce our food waste and our reliance on animal agriculture to avoid further climate change. (Refer to chapter 5 'Food' for more.) Also, some biomass has land-use implications if forests need to be cleared to make way for the crop. It also has the potential to emit amounts of CO_2 that make it comparable to natural gas if not done sustainably. So, not all green gas is equal!

Nonetheless, there are "options that can play a part in meeting energy needs and in balancing variable green electricity power sources."[93] Synthetic gas production from wind farms is one such promising proposal to decarbonise. Synthetic gas is a natural gas alternative, where hydrogen is made using surplus power from a source – such as a wind turbine – to create gas through electrolysis.

But government organisations, including Resources for the Future (RFF) and the World Energy Council (WEC), are not taking sustainable bio-gas opportunities like these ones on-board. Instead, their scenarios incorporate continued fossil fuel use to meet energy needs. This might seem surprising until one considers that the government's role is to stay out of the way of businesses' profit-making plans within our 'free market' system. And, under the umbrella of 'look-after-number-1' capitalism, there's an impetus for politicians to benefit by appeasing the wishes of big business. This is thanks to lobbying, including gift-giving to politicians, and revolving employment doors, or "The movement of high-level employees from public-sector jobs to private-sector jobs and vice versa."[94]

Even if not surprising, it's highly troubling that our future currently lies in the hands of governments because all the trappings of the CC22 seem to be preventing them from putting the necessary policies in place. That's why our vote for a safer climate and better life using our wallets is so important, and we should take any opportunity we can to invest in our future.

One UK supplier called Ecotricity has plans for a grass-powered, bio-gas plant. This vegan energy source is touted on their website as being sustainable as it doesn't emit *any* CO_2 and it won't run out. Apparently, it has several benefits:

- It's grown on land previously used for grazing livestock.
- It allows wildflowers to grow, which attracts pollinators and preserves biodiversity.
- The co-product of production is organic fertiliser.
- It improves soil quality allowing for successful food crops to be grown in rotation.

On top of all that, it provides carbon storage in the soil as it draws down carbon from the atmosphere.[95]

As great as it sounds, I'm sure there would be downsides to producing grass-powered green gas on a global scale. But, for now, it's a huge improvement on burning fossil fuels. And while we have gas powering our homes, we have to support more sustainable options.

With that in mind, I decided to change my supplier to Ecotricity. Until I read up about this, I was happy with my renewable electricity supplier, and my plan was to move away from using gas to power my home at the first opportunity. Since I found out there's the chance to support a sustainable source of green gas, I'm going to help ensure it gets the investment it deserves.

Energy-efficient home - retrofitting

Retrofitting has more appeal than just the way it sounds. Retrofitting your home involves making modifications that improve energy efficiency and decrease energy demand, thereby reducing your carbon footprint, your energy bills, and increasing your quality of life.

To help us reach our 0.8 tonnes of carbon savings, there are some easy fixes, which lead to immediate savings on energy bills. There are also some more significant home improvements. These cost more upfront, but have a return on investment both financially and for improved wellbeing. Indeed, research suggests that keeping our homes well-insulated and draft-proof has both physical and mental health benefits, including reduced depression and anxiety.[96]

I used a medium-sized home as an example for this chapter because I'm assuming most of you are not multi-millionaires or billionaires with humungous houses. However, for anyone that does sit in that bracket, or even a notch or two below it, the good news is that you'll be the first to afford the retrofitting measures that can make your home zero carbon, or even regenerative. That means that instead of creating the lion's share of emissions from energy use at home, you could be part of the solution by producing and exporting energy!

So, depending on your personal situation, it may make sense for you to install new systems and appliances, and to fully insulate your home straight away. Or, it might be more practical for you to plan them in as part of your home refurbishment plan over the next several years.

It might seem tempting to bulldoze old houses, and start again, using all the modern design features for a regenerative home, but it's more energy-efficient to retrofit our homes than to knock them down and build new ones.[97]

So, from wall insulation to draft excluders and sexy systems (see the section 'Systems thinking' below) to careful use, there are things we can all do to make our homes more comfortable as well as climate-friendly.

Lighten up

Turning the lights off at home when they're not needed is a bit of a cliché. Having said that I was surprised to learn from ethicalconsumer.org that lighting accounts for 18% of household energy use. Also, that LED bulbs can be up to 90% more energy-efficient than incandescent bulbs. To lighten up the carbon load that your bulbs produce, there's a handy guide to buying light bulbs at ethicalconsumer.org.[98] So, just by getting into good habits of turning lights off, and using the most energy-efficient bulbs, you can knock a fair bit of carbon off your total. And save some cash. Another super quick and easy fix.

Insulation

Insulation is fundamental to making your home not only environmentally sustainable but also comfortable and healthy to live in. There are lots of fantastic new inventions and design principles that supplement well-insulated homes. They can give housing a small carbon footprint, or even contribute to an overall reduction in emissions, e.g., regenerative housing. However, without the foundation of efficient insulation to prevent the loss of heat or cool air from your home, you'll be fighting a losing battle.

Half of all your heat can escape if your home is poorly insulated.[99] A quarter of that heat is lost through your roof.[100] This means that if your roof isn't properly insulated, you should make this your first job. It will save excess carbon production as well as making an immediate impact on your heating bills. Happily, it's also cheap to buy a natural, recyclable material for the job such as fibreglass or mineral wool. On a detached house in the UK, the installation would cost you an estimated £395, but you'd make back £225 – over half(!) – in energy bill savings in just one year.[101]

There are other insulation jobs around the home that are inexpensive and make sense to do straight away.

These include:

1. Fitting draft excluders to external doors.
2. Repairing gaps/resealing window frames and replacing 'blown' double glazing (this happens when condensation gathers in-between the windowpanes when the seal fails).
3. Window dressings like thick curtains or blinds with thermal backing.
4. Filling any gaps in the floor or using carpet/rugs.
5. Covering your hot water tank (if you have one) with a specialised jacket.

Here are some things that you could either do immediately, or put on the plan for future home improvements:

1. Effective double or triple glazing.
2. Internal wall insulation for cold walls (drywall).
3. External wall insulation.

I won't go into detail on any of these here, but if you want to install them, shop around for a reputable company as well as the best deal.

I can personally vouch for making a number of these changes. Implementing these improvements made a huge difference to the comfort of our home and our quality of life. Improved acoustics was also an unexpected benefit! "Replacement windows, external doors and new façades and insulation can reduce external noise transfer into the home, helping to contribute to residents' wellbeing."[102]

A WHO study found that in Western Europe, at least a million 'healthy-life' years are lost each year due to exposure to environmental noise. After air pollution, this makes noise the second largest environmental cause of ill health.[103] If you've ever had traffic or building work noise where you live, I'm sure you can relate. In our last property, we had what I can only hope was someone too young to know better using our block as a motorbike race circuit. Until we replaced old windows, we weren't able to do much relaxing.

Temperature control – passive or massive?

What struck me most when researching for this chapter is the number of environmentally-considerate options that are now available for controlling the temperatures in our homes. And, as I already mentioned, once your home is well-insulated, you can really begin to benefit.

Energy smart

The easiest and cheapest thing to install, first, is a smart meter. The premise is that if you install one, you can closely monitor and control your energy (gas and electricity) so that you can reduce your usage and save on bills. This is because you get accurate and real-time info on your energy usage, which will encourage you to think about where you might be being wasteful. You can also control your heating from an app, so if you're having too much fun in the pub to go home, you can delay your heating coming on.

A smart meter could help reduce your electricity use by 2.8% and gas use by 2%.[104]

If you don't have a meter, you should still consider where you might be overusing gas. Do you have the heating on in rooms that you're not using? You'll save money and energy by turning it off (if you never use the room, then make sure you still air it out, and warm it up at times, to avoid damp).

Turning radiators off at the valve, when not in use, is in the top six energy-saving behaviours in households, according to Cambridge Architectural Research. The top energy-saving behaviour in households is turning the thermostat down by 1 or 2°. So if you can live with reducing the temperature by a degree or two, and wearing warmer clothes instead, you'll be making the most impact.[105]

There are multiple benefits to reducing usage. Most people know that extracting gas and burning it to produce electricity creates air pollution and greenhouse gas emissions. But, you might be surprised to learn that emissions from gas heating and cooking contribute to 20% of Nitrogen Oxide (NOx) pollution in urban areas, which causes multiple health problems.[106]

Systems thinking

Maybe you have a larger-than-average house with a subsequently larger carbon footprint, or you're just keen to make your property energy self-sufficient or regenerative. If so, then you can take steps that require a bit more investment and planning.

You might want to consider making your home airtight and installing a heat recovery ventilation system. In many properties, airbricks are required to allow ventilation so that you can avoid damp, stale air, and unhealthy mould-growing conditions. The problem is that these holes in our walls let cold air in and let heat escape, which is not conducive to low energy usage.

A heat recovery ventilation system could be the answer. It consists of a unit that is connected to pipes and valves that extract damp, stale air from each room. It then recovers the heat from that extracted air and uses that heat to warm fresh air from outside and supplies that back into the house.[107]

This type of system reduces the need for traditional heating systems and therefore moves your property on the dial towards being passive rather than active for energy use. Housing design can incorporate truly passive systems that require no energy to run, such as using direct sunlight to heat your property. However, these types of systems are more effective when designing and building a new house. If you happen to be building a house and want more information on passive systems and other aspects of regenerative house design, such as using carbon-sequestering materials, then one useful link I found is:

https://sustainability.williams.edu/category/green-building-basics

Carbon negative = climate positive

If we're retrofitting, we may not be able to install heating and cooling systems that are entirely passive. However, there are temperature control systems that are far less energy-intensive than traditional central heating and air-conditioning. These can be combined with renewable energy generation and storage (using batteries) at home so that your property can generate more energy than it uses and export the rest.

Systems thinking can = system change

As mentioned, in the UK, heat pumps were included in most of the key political parties' manifestos on climate.

Heat pumps are devices that come in three forms: ground source, air source, and water source. Simply put, they transfer heat from one source, (e.g., from under the ground if it's a ground source heat pump), to another location (e.g., the water in your radiators). Many of them can also reverse the heat collection process to provide cooling when required.

Another option for heating your home with a low carbon impact is to install infrared heating panels. These have been around in Europe for a while, but are relatively new in the UK. This exciting technology has many benefits compared with traditional heating systems. These include aesthetically-pleasing heating panels, creating a healthier home by preventing damp and mould, highly-efficient energy use which makes the running costs comparable to gas central heating, plus low maintenance and easy installation as there's no boiler or pipework required. Also, "Studies have shown that infrared heating technology is the most effective way of providing comforting heat for humans which is the reason why it is used in incubation units for babies."[108]

Investments in energy-efficient technologies like heat pumps or infrared heaters, combined with renewable energy generation, will often result in an eventual financial return on investment. But they are about so much more than finances – they are an investment in our collective future. The more we make retrofitting mainstream by voting for it with our wallets, the more likely that this culture shift will be followed by more government investment in retrofitting for those who can't afford it.

What's more, many forward-thinking economists are telling us that government spending on green energy and retrofitting is better for a healthy economy than consumer spending. "For every £1 invested in retrofit, an estimated £1.27 would be returned in tax revenues". And, "23 person years of employment could be created for every £1 million invested in retrofit."[109]

"More money spent on retrofit means less money spent on energy – investment, rather than consumption spending. It is argued that this change in spending patterns will result in higher tax intake for government via higher net levels of employment (UKERC, 2014) and improved fiscal returns."[110]

In his book "Prosperity without Growth", Tim Jackson argues that this kind of spending can kick-start a non-growth economy without incurring job losses. A non-growth economy lends itself to other success measures that consider overall wellbeing, unlike the monetary GDP goal by which we currently measure ourselves. And, it is what is required for a long-term sustainable future as humans on this planet.

In the UK, we could save £8.61 billion in energy bills every year if homes were improved to Energy Performance Certificate (EPC) rating C. The ratings start at A! So, even if only UK citizens make housing that is really rubbish at energy efficiency a bit *less* rubbish, we're talking billions less in energy bills. Saving the planet, and saving money, too.

Generate to be regenerative, and help create a positive future for all

Heat pumps can be combined with renewable energy generators such as solar panels and/or wind turbines. Most of you will be aware what solar panels are, although perhaps less aware that they can be used to heat hot water as well as create electricity. I'm sure you'll also be aware of wind turbines, but possibly not have thought whether it is possible to install one at home. If you want to find out more about installing these systems, it's definitely worth putting a chunk of time into research. It's beyond the scope of this book, but you'll need to be sure that what you install will create the most energy efficiency, and generate the most renewable energy for your specific home. Always ensure a reputable company does the installation and properly advises you about ongoing maintenance.

A combination of the measures in this chapter can reduce your energy demand and potentially lead to you producing more energy than you need. This excess can now be sold back to energy suppliers in the UK. (Check online for schemes in your country.) Going forward, this local energy can be distributed amongst your community from a local smart grid.

Thanks to modern technology – from generating renewable energy at home to 5G capability for the Internet of Things (IOT) – local smart grids are now emerging. These local electricity networks monitor and integrate the electricity supply and demand of all connected users. In doing so, they ensure the system is economically efficient, sustainable, and offers a secure and safe supply of energy.

Installing renewable energy at home means that – as these grids develop over the next few years – the excess energy you produce can be efficiently distributed within your local community. The community as a whole can benefit. Alongside retrofitting, this can completely eradicate the need for a national grid supplied by fossil fuels.

Importantly, you'll also be supporting the retrofit and renewable energy uprising. This means you'll be giving the government a clear message. That you're investing in a positive future before it's too late, and now they need to do the same.

Key takeaways from this chapter

- 20% of all carbon impacts are down to how we fuel our cars and homes. Fossil fuels have propelled us, as a species, to the high standard of living that many of us experience today. But as they're running out, so is the planet's capacity to maintain a stable climate.
- We must transition to renewables. Many academics and NGOs have created plans to do this that will also have a positive impact on the economy and our health.
- It might be more difficult to understand why these plans are not being implemented as quickly as they need to be, if we hadn't explored the CC22 root cause.
- Our governments are not acting fast enough to avoid climate catastrophe, so it's up to us to give businesses the steer they need to speed up the transition to renewables.
- When talking about electricity, there is a very easy fix – switch to a renewable supplier. Or install your own means to create renewable energy at home.
- You can also move to a more sustainable gas supplier with a bit of research. But the most effective solutions to move away from gas come from retrofitting your home. After you've insulated your home, you can reduce your emissions further with careful usage.
- If you're keen to reduce emissions to a minimum and maximise the surplus energy you generate, you can combine creating renewable energy at home with some longer-term investments in technology like heat pumps and heat recovery ventilation systems.
- In doing this, you'll not only be ready to share surplus renewable energy with your neighbours, as 5G Smart Grids emerge, but you'll send a clear message to businesses and governments that you support the transition to a green economy.

7 | Recycling

Initially, I wasn't going to write about recycling in a standalone chapter. Because, as we're about to see, it should be a last resort. And I didn't expect the carbon saving from recycling to be especially big. But, according to footprintcalculator.org, if you go from never recycling to always recycling your paper and plastic, you could save a massive 1.7 tonnes of carbon per year.

The other reason that recycling began to seem deserving of more keyboard action is because of all the confusion around the subject. Recycling is often in the mainstream media, but that only seems to add to the uncertainty around whether it's worth doing, how to do it, what can and can't be recycled, and whether it might even be making things worse rather than better! This chapter attempts to answer these questions, plus find out which common recycling myths can be dispelled.

Recycle last

When people hear the moto 'reduce, reuse, recycle', it seems like the only word people actually hear amongst those three is 'recycle'. But recycling should be – as implied by the phrase itself – the final option. If you've caught it on the news or social media, you'll be aware that there are issues with recycling. Even if you play your part, some of the processes are not 'fit for purpose', which in reality means that not everything gets recycled, and much of it is exported to poorer countries to be sifted. What doesn't make the grade is incinerated or just dumped in illegal landfill sites, and the people who receive and deal with it are exposed to severe health issues, as is the surrounding land and wildlife.

For some, this news is a reason not to bother with recycling. It makes it easy to brush it off with a 'What's the point?' attitude.

On the surface, that seems fair enough. But anyone that has given it a little more thought will realise that we can't just ignore it, or give up, if we want a positive future for ourselves and our loved ones. Leaving a toxic and carbon-emitting planetary junkyard as a legacy is simply not an option.

Some people may believe that the only alternative is to try to force policy change; that it's all up to government. Surely there's nothing we can do about our recycling being illegally dumped abroad? But recycling is a perfect example of where policy alone fails without the actions of the individual to back it up.

In many developed countries, policy for recycling has been in place for decades. But the central reason that things have got so messed up is because we individuals, acting within the current system, have done very little to reduce and reuse. Or change anything about our throw-away lifestyles.

We, in our limited knowledge, assumed that recycling (e.g., chucking everything into the bin on the left, instead of the bin on the right) would solve all of our waste issues. It's pretty clear now that the *amount* of stuff we buy, and ultimately chuck in one bin or another, is the real issue that needs resolving, because no amount of recycling is going to make the problem go away; we just can't keep up.

This means certain policies on waste intended to improve the situation (such as charging to put things in landfill) can actually encourage the wrong behaviours – like illegal dumping. As with all climate change issues, policy alone cannot resolve it because it's not dealing with the root cause of the problem.

Reduced, reused, now recycle

Each chapter in this book has talked about the products and services we buy. In each one, there's guidance on minimising our purchases or swapping them for something less damaging. Assuming you're adopting these suggestions, we don't need to go into more detail on the 'reduce' and 'reuse' part of the moto here. You're already on your way to taking personal responsibility for preventing catastrophic climate change.

Recycling was always meant to happen after reducing and reusing has been considered. And when thought of in that way, it definitely isn't a waste of time. But recycling will be necessary until we have collectively shifted culture to this mindset, and subsequently shaped business and government policy. So the next few paragraphs cover how to go about doing it effectively.

Government guidance on how it all works may feel unclear, but we have the opportunity to educate ourselves on how to go about it. Then we give the stuff we've used the best chance of finding new life instead of rotting away. In this chapter, I also hope you'll gain inspiration from some of the fantastic recycling initiatives that are happening around the world that go beyond government policy and our 'green' bins.

Not where we need to be

You probably do your best to recycle. But the feeling of hopelessness (from not knowing the difference it can make), and a lack of guidance, may be holding you back from doing more, or doing it properly. I know this because we're all in the same boat and, until researching for this book, *I* hadn't been doing the best that *I* could. Now that I've got the lowdown and put some improvements in place, I can help to make it easy for you to do too.

To line ourselves up for success before attempting the solutions, it helps to deal with the blockers that have been stopping us from getting going. That's why the breakout box highlights why recycling is not a waste of our time.

What a mess

If the UK is to reach its target of net-zero emissions by 2050, then it should place a ban on biodegradable waste being sent to landfill by 2025.[111]

In other words, the amount of CO_2 and Methane (a much more potent greenhouse gas) that's released from landfill is significantly contributing to climate change. That's not the only issue with landfill, of course. They are toxic dumping grounds of anything we discard that quite literally leaches into the surrounding land, groundwater, and waterways.

Unfortunately, it's a case of 'out of sight, out of mind', because if we could see or smell the mammoth mounds, we may be less likely to want to add to them. But add to them we do. In 2016, each EU inhabitant generated 5 tonnes of waste; that's an almost unimaginable amount, and the majority of it is landfilled. As well as all that, not recycling means using virgin materials each time something is produced. To give you a (bad) taste of what that means, every tonne of non-recycled paper means cutting down 17 trees, taking up 3.3 cubic yards of landfill space, using 70 percent more energy in production, and increasing air pollutants by around 200,000 tonnes.[112]

The official EU statistic is that 37.8 percent of materials were recycled in 2016, but as far as I can tell that doesn't take into account the stuff that gets shipped abroad and dumped because it's so contaminated it can't actually be recycled. The negative press around this, and general confusion around recycling, is making people think that it's a bit hopeless. But the beauty of having our clear goals and measures is that we can assess how much impact we're having on climate change by recycling, and I'm happy to say it's a big-un.

"Recycling… has been shown to save over 700 million tonnes in CO2 emissions every year… Not only that but approximately 1.6 million people worldwide are employed in processing recyclables. The annual contribution of the recycling industry towards the global GDP is projected to exceed US $400 billion in the next ten years. $20 million dollars is invested each year by the industry into job creation, improving recycling efficiency and environmental impact."[113]

Don't waste your consumer power

Recycling and purchasing recycled products is another way we can – with our consumer hats on – drive positive change in the system. Waste companies do their best to recycle your waste wherever possible, and recyclables are considered a commodity because consumers like products made from recycled materials. This increases their desirability to manufacturers who then buy more recycled materials for their products. The prices for the materials then increase, "which means recycling programs remain feasible."[114]

Waste companies are even charged fees to dump waste at landfills so, ultimately, it makes no financial sense for them to do it. It's reassuring to know that if we do our bit properly, then the waste companies will do theirs.

We know that the ecological footprint calculator gives 1.7 tonnes of carbon saving per year by going from someone who never/rarely recycles to someone who recycles virtually everything, (i.e., zero(-ish) waste). Before doing the research, I would have guessed that being zero waste is virtually impossible. And to practise an entirely zero-waste lifestyle would be very tricky indeed, but to be zero-waste-ish is actually doable.

This is because there are recycling schemes outside of the home that mean virtually everything has a place somewhere. Don't get me wrong, you're not going to be perfect, and you don't have to be. There will always be some stuff that ends up in the bin for one reason or another, but you will be able to achieve a big chunk of the potential carbon footprint saving with some simple improvements.

All it takes is a little organisation, some Lean/Agile tricks, and some guidance for us to mobilise as an army of nearly-zero-wasters. Easy. And recycling instead of putting things in landfill means more than just your conscience will be clean. You'll be someone who's making sure our collective home remains liveable in, for ourselves and the ones we care about the most.

Recycling solutions

It's important to note that we'll talk about the various materials that can be recycled in turn. When looking at the production to the decomposition of products, they have hugely different impacts and outcomes on climate and the environment. Bulking all recycling into the same proverbial boat is unfair when talking about what is shipped out to poor countries and ends up back in the ocean, because this is largely plastic. This means there isn't one solution for all recycling. Instead, there are multiple small improvements that help ensure each material causes the least amount of environmental damage.

Plastic

Note that in the rest of this book, the biggest carbon footprint saving solutions come first. But as it is such a huge topic right now, I'm first going to do what nature can't, and perform a quick breakdown on plastic.

Although plastic waste is at the forefront of everyone's mind when it comes to the environment, it's not the most important material to recycle to avoid climate change. It's a perfect example of how lacking specific goals and measures for making improvements to tackle climate change can send us down a reasoning rabbit hole.

Yes, plastic *is* causing an environmental crisis. But because no formal problem-solving techniques – such as Lean/Agile – are being used to solve it, the solutions are often not suited to the cause.

The two problems most often talked about in relation to the plastic issue are:

1. It takes many (often hundreds) of years for plastic to break down.
2. It's clogging up the oceans and killing marine life.

This is creating a multitude of businesses looking to swap plastic for other materials, as it seems people are desperate to avoid it, without much thought as to the potential consequences. Businesses that now market their products and/or packaging as biodegradable have got a big 'green' tick in consumers' minds right now. Surely something that breaks down is better than something that hangs around for hundreds of years? With no measures in place for the public to utilise, things get confusing. So, let's be really clear – when talking about whether something that breaks down is *better* than something that hangs around for hundreds of years, in terms of aggravating climate change, the answer is a resounding NO.

Climate change is caused by the release of excess greenhouse gasses into the atmosphere. A proportion of these gasses come from when something breaks down, and this is maximised in the conditions of landfill. Even something as seemingly 'green' as compostable packaging releases harmful gases in a landfill environment.

It's worth pointing out here that most compostable packaging in the UK can only be composted at specialist plants, not in your food bin. This means that putting compostable packaging in with your food bin will contaminate it and possibly mean the whole lot being sent to landfill where it will release the potent greenhouse gas – methane.

As plastic generally doesn't break down for a long time, greenhouse gases are not being released when it's discarded. "Inorganic carbon in plastics does not decompose through the anaerobic digestion processes, which results in no emissions if it is landfilled."[115]

This means that *waste plastic is not a significant contributor to climate change.*

The other thing to call out is that plastic generally does not end up in the ocean in countries with waste management systems in place, (i.e., developed countries). As long as you throw it in the bin, you can be fairly confident it will end up in landfill. However, the majority of recycled plastic is exported to poorer countries that don't have waste management systems. And, as we know, there's so much of it – and it's sometimes so contaminated – it cannot be recycled. This means that in first world countries, *the plastic you recycle is more likely to end up in the ocean.*

All this begs the question – should we recycle plastic at all?

It's a valid question when considering the above points, but the simple answer is yes. And now we have a better understanding of the process, which Lean teaches is vital to problem-solving, we are able to put effective solutions in place.

The allure of not having to bother is tempting, but the alternative of letting mountains of toxic plastic build-up for future generations to worry about isn't a valid solution. Another reason it's worth doing is because virgin materials use more resources and energy and therefore release more greenhouse gasses in *production* than recycled plastic. So when it works, the plastic you recycle prevents more fossil fuels being used to produce and refine virgin plastic. The great news is there are some inspiring schemes to get involved in for recycling plastic now. These make the most of the ingenious properties of plastic rather than dumping it somewhere where it's no use to anyone now, and will cause everyone a headache down the road.

Recycling plastic - solutions

Continue with kerbside recycling (if you have it in your area).

The risk with this, as we've said, is that it could well end up getting shipped abroad (two-thirds of UK plastic is) and possibly not recycled at all. The way you can minimise this risk is by ensuring that your recycling is not contaminated. Contamination happens either by placing non-recyclable materials in with your recycling, or by not rinsing leftover substances.

Plastic comes in many forms, and not all are recyclable from kerbside collections. The way that you can normally tell if it is, is by the number printed inside a triangle of arrows somewhere on the packaging. It's handy to have an idea what these numbers mean. Some packaging can even save you unnecessary purchases of reusable containers. If it has a number '5' printed in the triangle, it means it's both reusable *and* recyclable. The plastic pots that takeaway food comes in are a great example. I've found them super handy for storing leftover food in, time and time again, in order to keep it fresh and help prevent food waste as well.

Exactly what plastic can be recycled from your kerbside depends on the area you're in, so you'll need to do a quick check before you get started. In the UK, you can go to https://www.gov.uk/recycling-collections and enter your postcode, which will then redirect you to the relevant website for your area. If you're based in the US you can go to:

https://how2recycle.info/check-locally.

Always *check* the packaging, as brands will differ. But this chart gives you an idea of what the numbers mean in the UK. A quick internet search should give you this info if you're based elsewhere.

	Used in	Once recycled	Recyclable at home (check your area on the council website)
1 PETE	Food packaging such as punnets and soft drink bottles	Used to make more PET products	Usually
2 HDPE	Yoghurt pots, cleaning products, milk bottles, shampoo	Used to make garden furniture, pipes, milk bottles	Often
3 PVC	Window fittings. Thermal insulation, kids toys	NA	Never
4 LDPE	Shopping bags, magazine wrapping	Used to make bin liners, plastic furniture and floor tiles	These are normally collected at supermarkets
5 PP	Ready meal trays, margarine and yoghurt tubs, takeaway containers	Used to make clothing fibres, food containers, speed bumps	Sometimes
6 PS	Plastic cutlery, disposable coffee cups	Used to make more packaging	Sometimes
7 OTHER	Salad bags, vegetable packaging, crisp packets	NA	Never

What has now been revealed in the news (in case you missed it), is that we've been duped for years into thinking that the symbol with two arrows that swirl around and interlock inside a circle – often found on crisp and chocolate packets – means something is recyclable. It doesn't. All it means is that the company has donated some money towards recycling.

If this is the first time you've heard this, you might be feeling pretty cheated. I know I was. I felt conned and annoyed at all the recycling I would have contaminated by throwing those sorts of wrappers into the recycling bin. However, there are now schemes where previously non-recyclable plastic can be put to good use, which I'll go into in the next solution option.

Use an alternative scheme

I'm going to name a couple here that are multinational, but please take a look to see what else is going on in your area, in case there's another one that suits you better.

TerraCycle

Conscientious companies fund multinational TerraCycle (TerraCycle.com), which means it can offer a range of free recycling programs for hard to recycle waste.

It's aiming to eliminate the idea of waste altogether by turning old materials/products into new, useful ones. Not only does TerraCycle not landfill anything, it also doesn't incinerate any waste, as this does not make use of the original materials and the process can release harmful gasses into the atmosphere if they're not captured correctly.

TerraCycle offers several schemes for nearly every type of plastic you can think of that clogs up your bin or recycling box. From confectionary packets and baby food pouches to pet food packets, personal hygiene packaging, toothpaste tubes, and toothbrushes. Take a look at the available schemes on the website and find out what else you can recycle and where your nearest collection point is, (or you may be able to set one up yourself from home, school, or work). If you do, you are rewarded with points, which can be exchanged for donations to a school or charity of your choice.

Taking part in a TerraCycle scheme should tick some of your wellbeing boxes, too. You'll get benefits from improving your natural surroundings and creating a positive future for yourself and loved ones. The scheme I participated in (before lockdown) was held at the local library where all sorts of 'non-recyclables' got collected. Going there had a refreshing community feel and an all-round feel-good factor. Setting up a collection point has the added bonus of giving to charity or raising funds for your school to stretch the satisfaction even further.

Ecobricks

Ecobricks gives you an alternative, equally brilliant option for making sure that your non-recyclable plastic is put to good problem-solving use instead of causing problems later on. The idea is to fill a clean, empty plastic bottle full of clean, plastic non-recyclables such as crisp packets and fruit and veg packaging. Yes folks, while we wait for supermarkets to ditch unnecessary plastic wrap, that pesky stuff that grates your nerves – like the carrots it contains – can be turned into something positive.

Ecobricks take the properties of plastic currently thought of as negative, such as how long it takes to break down, and utilises them as positives. Plastic makes a really durable building material, so the bricks are being used both in the UK and abroad to build things like playgrounds and furniture and even entire buildings.

You can sign up to view, join, or create a local community, and get in touch to see where you can drop your Ecobricks.[116] Just make sure you follow the straightforward guide to ensure your bricks can be used.

Taking part in this initiative also has a feel-good factor, and if you have kids, it's a family activity you can do to teach them about environmental responsibility and how waste can be used to better our communities. If you're a teacher reading this, or a parent who would like their kids' school to be involved, it would make a fantastic project with lots of learning points, from saving landfill space to sustainable building solutions in developing countries.

A combination of the two

It may be easiest and most practical for you to combine kerbside recycling with other options. If you choose to do this, when you kerbside recycle, give the packaging a rinse before chucking it in the recycling box (making sure you follow any specific rules of your local area to avoid contamination). For anything that isn't recyclable by the kerbside, or if you're not sure because there's no recycling number to go by, or it's got a symbol or number you haven't seen before, then put it in with your chosen alternative recycling scheme.

Other materials

This is where recycling becomes a no-brainer because of:

- The ease of recycling materials other than plastic.
- The success rate of recycling materials other than plastic.
- The benefits it brings in preventing greenhouse gas emissions from landfill.
- The CO_2 emissions and resource-saving in production compared with virgin materials.

Because most materials other than plastic are much easier and simpler to recycle, it often happens nationally. Plastic, unfortunately, just gives recycling a bad rap! Below is a quick rundown on the main materials in order of their biodegradability (i.e., greenhouse gas emissions impact), and what to do with them and how.

Food

If you're like the average person from a wealthy country, you'll throw away 30-40 percent of the food you buy.[117] [118]

In the chapter on food, we explored the ways we can minimise this waste because prevention is always better than the cure. However, at least some food waste is unavoidable, and what happens to that waste has the biggest and most immediate effect on climate change out of any of the other materials.

This is because, in a landfill environment, the lack of oxygen turns the food waste into methane, which (as mentioned) is a 25 x more potent greenhouse gas than carbon dioxide (when measuring its 100-year global warming potential).[119] It also happens very quickly compared to the breakdown of other materials in landfill, and because there is so much of it, this adds up to a big problem.

You can overcome the issue of food waste by starting a home compost. I won't go into the *how* here, because an internet search will give you much better advice on your *specific requirements* than I can. It's worth mentioning again that you *can* home compost compostable packaging... so you can kill two birds with one stone there. You'll also help keep your garden in tiptop shape without nasty chemical fertilisers, so three birds would cop it... as per the expression. In reality, of course, you're also doing your garden wildlife a favour.

The other/combined option is to dispose of food waste using the local authority food waste collection service. They will either recycle it into a good quality fertiliser or use anaerobic digestion to break down the waste, collect the methane, and turn it into biogas that is used to generate electricity instead of relying on fossil fuels.

To find out more, visit recyclenow.com

But, before you go any further, a point to investigate is what type of 'bin liner' your authority accepts. Some authorities won't accept liners such as plastic bags that don't break down, and using them could mean the lot has to be redirected to landfill. I've discovered that in the UK, you can get hold of innocuous food waste bags from your local library that are made from potato starch for free!

If you live outside the UK, or don't have a food waste collection service near you, there may be a food waste recycling centre you can drop your food waste off at instead.

Paper

Paper is similar to food waste in landfill; the lack of oxygen means that when it breaks down, it also produces methane. It may take a few years to break down, so the effect isn't quite as immediate as food but, in terms of the effect on climate change within our lifetimes, the outcome will be pretty similar. Therefore, as paper is so widely and easily recycled, it's most definitely worth doing. I expect that you already are but perhaps, like me before doing this research, there may be some simple rules you're overlooking that are diverting your well-intentioned recycling attempts to landfill.

We've talked about what happens to contaminated plastic, but if any type of material placed in recycling is too contaminated to process, it will also end up either in landfill or being incinerated, releasing harmful greenhouse gasses.

"Since your recyclables will eventually be sold to manufacturers, they must meet certain standards. They can't have too many impurities, since recycled materials compete with virgin materials for use in manufacturing. So the cleaner the materials you return, the more likely it is they will be recycled... a pizza box covered in grease and cheese you toss in your recycling bin will end up in a landfill."[120]

A few easy tips from the recyclenow.com website:

- If you scrunch paper and it doesn't spring back, then it can be recycled.
- Remove sleeves, (e.g., a paper sleeve that is wrapped around a large yoghurt pot can be removed and recycled with paper). However, leave labels on.

- Leave lids on food and drink cartons as that will mean they get recycled too, instead of falling out of the process with other contaminants.
- Empty and rinse everything that's washable.

In the UK, check recyclenow.com for extra guidance; you can enter your postcode to find out what type of paper and card your area accepts.

Steel cans

"On average, your household gets through 600 steel cans every year... about two per day."[121] Cans containing things like baked beans, biscuits, paint and aerosols are easily recycled and although Britain uses around 12.5 billion steel cans every year, we only recycle about half of them. They're 100 percent recyclable an infinite number of times, and can be used for virtually any process, like the building of new cars. "You could actually be driving about in something that was once a baked bean tin."[122]

Steel takes about 50 years to degrade, so it doesn't have quite such an immediate benefit in terms of landfill gas release as food and paper. But, considering all you have to do is give it a quick rinse and chuck it in the kerbside recycling bin (in the UK and many other places) to make use of it over and over again, it's a bit bonkers to bin it.

Aluminium

Aluminium is one of the easiest and fastest materials to recycle, and aluminium cans can be recycled and reused within as little as 60 days. Also, recycling aluminium saves almost 100 times the power than preparing aluminium products from virgin metal. It does take about 250 years to degrade, so it's not letting off greenhouse gas immediately if you chuck it. But, with that said, filling up landfill with drinks cans – when it's so easy to stick them in the recycling – is beyond rubbish.

Glass

According to All-recycling-facts.com, glass is one of the few products that can be completely recycled over and over again. When it ends up in landfill, it never decomposes.

We're in a similar boat to plastic here, with regards to landfill emissions, but, "There is an increased demand for recycled glass referred to as "cullet" in the glass industry. This is because cullet or recycled glass costs much less than raw materials used to manufacture glass from scratch. Cullet also consumes very little electricity."[123]

To find out how to implement these suggestions easily using Lean/Agile, refer to the Appendix sections 'Continuous Improvement', 'Scrum – Sprinting', and 'The 5S System'.

Everything else

Hopefully, you've been implementing the improvements in this book in order of the highest carbon footprint savings for you personally, as per your results on footprintcalculator.org. If recycling is the next thing on your list and you've already improved your recycling habits on the key materials above for the biggest carbon footprint gains, you may want to move on to the less-frequently recycled items. You can use the guidance below to assist you.

There is a guide on the Bournemouth Council website with an A-Z of recycling which is an extensive list of what to do with every manner of item. It's written by my local council but is mostly relevant to all of the UK and beyond, and it includes *reduce* and *reuse* advice. For example, if you absolutely have to use clingfilm, then you can recycle it with carrier bags in your local supermarket's plastic bags recycling bank. Also, old bedding can be given to pet shelters. Nice! Alternatively, you can use a local guide you locate online.[124]

Getting going

Your mind might be boggling at the thought of being organised enough to recycle even the main materials you use every day, so don't forget to refer to the Lean/Agile guidance in the Appendix.

In the case of recycling, it's mainly about getting set up for success. Once everything is in its place, and you're up and running, it's just part of the daily/weekly/monthly routine. The difference being that this new routine is the routine of someone you'd rather be – a person who takes more social and environmental responsibility and can rest easier at night for it.

Key takeaways from this chapter

- There's a lot of what seems like conflicting information about recycling that can lead us to question whether it's worth doing at all.

- It's important that we use recycling as a last resort after we've reduced our usage and reused what we can. When thought of this way, recycling is an integral part of our impact on the climate and can give us a carbon footprint reduction of 1.7 tonnes per year.

- Recycling is another area where we can use our consumer power. The more recycled products we buy, the more demand there is for recyclable materials from waste companies who turn them into something new, rather than letting them pile up in landfill.

- Plastic waste contributes little to climate change, but some of the biodegradable alternatives do, especially if they end up in landfill. So, be mindful of what you're replacing plastic with, and how you dispose of it.

- If you're unsure whether plastic is recyclable at home, don't put it in household recycling as contaminated plastic recycling is more likely to end up doing harm to the environment than if it's in landfill.

- You may want to recycle plastic carefully and thoughtfully in your household recycling, or you may want to use another scheme that can make good use of it. Or you may want to do both.

- Recycling other materials at home is an easy win as long as you get organised and do it properly. Visit the Appendix for some further guidance on this.

8 | Continuously Improving

I've deliberately avoided making this book an arbitrary list of things you can do to fight climate change. The two reasons for this are that there's already a tonne of ideas in circulation – both in books and online – and secondly (and more importantly), they can put your head into a spin.

You'll have seen a thousand posts and articles on reducing plastic waste, Meat-Free Mondays, packaging-free shops, zero-waste lifestyles, and the rest. Personally, when I went food shopping, my head would come close to exploding.

I'd be trying to work out whether I should get the local veg that was wrapped in plastic, or food that came from overseas without packaging. Whether I should avoid palm oil altogether (which counted out the vegetarian meals I picked up; yep, I can't cook), or whether I could live with myself for having some organic, free-range, local meat.

I always ended up in an eco-decision maze that left me feeling disheartened. If I hadn't thought to apply a Lean/Agile approach to the problem, it would have led to me getting a bit 'over it all'.

It's not surprising that someone who's interested in how to avoid climate change can get caught up in where to start, and ultimately become overwhelmed. That's why this book is a guide on what can make the *most* impact. It reveals what goal(s) we should aim for, in order to keep us focussed and motivated to get there.

Hopefully, by this point, you're in the zone of putting measurable, impactful improvements in place that will benefit your life and target climate safety. If we begin to meet the 3-tonne CO_2 goal collectively, the world will start to look very different, and the future will look far more positive. But the CO_2 measure, courtesy of a footprint calculator, is not currently perfect. There are lots of specific lifestyle choices not covered by a calculator, which make a difference to the climate and our wellbeing.

With that in mind, you might feel ready to do some of the 'smaller things' that have been bugging you because they just make good sense. Pleasingly, these changes have an impact and are worth doing, especially once you've begun tackling the big improvements.

"Progress cannot be generated when we are satisfied with existing situations." Taiichi Ohno, creator of the Toyota Production System, which inspired Lean Thinking.

Once you've started, feeling a niggle to keep getting better – bit by bit – is crucial to making progress.

I've talked about continuous improvement as a method to make this book's suggestions manageable. And the continuous improvement principle can also be applied to implementing more and more advancements towards reducing emissions.

This chapter aims to cover just a few of these, which have become apparent to me on my journey to the 3-tonne goal. These come under three headings:

1. In your home, and on your phone.
2. Travel.
3. Localisation.

1. In your home and on your phone

How to cut back on tat

I've heard thought leaders suggest that choosing *what you buy* is like voting for something because you're backing it, and I agree. But you're not just voting for something; you are financially supporting it, which means it will continue to be produced and consumed.

We are – sometimes inadvertently – financially backing the companies we buy from.

With this in mind, it's worth asking ourselves whether we want our hard-earned cash lining the pockets of shareholders of companies that add no real value to our lives. And in many cases exploit their workers as well as the natural world.

I used to walk around shops having a browse, and think 'hmmm – that thingy might look nice in my living room', or 'that decorative box would be handy to keep things in'. And I filled up my trolley with stuff and spent a wad of money each time. But, once my awareness of the CC22 grew, if I found myself having those thoughts, I began to remember to ask myself before I got to the counter – *do I really need this*? Only a moment's more thought made me realise I didn't, and I was getting sucked in.

Now, the old habits have died – hard; it doesn't even occur to me to peruse the shops. And, I generally only buy things after I've spent some time considering the impact on my wellbeing and that of the planet. It's become completely instinctive.

I'm not telling you this to blow my own trumpet. I want to demonstrate that it's a case of building new patterns and pathways in your brain until they're second-nature. Then you'll be living mindfully, rather than in the haze of mindless consumerism that most of us are used to. At this point, it's much easier to make choices that avoid negative impacts.

"We are what we repeatedly do. Excellence, then, is not an act, but habit." Aristotle

Eco-friendly everything

When you do feel the need to spend your money on something, there are now, fortunately, more and more opportunities to give our cash to organisations that demonstrate that they share our core values by how they operate. Organisations that care about our wellbeing, and the wellbeing of the natural world.

Aim to build the habit of searching for the 'eco-friendly' options before pressing any Buy Button. Every time you swap out an everyday purchase for something that's more thoughtful towards the world around you, you'll also swap a bit of the meaningless feeling we (at least sometimes) get. This is replaced with a feeling that you're living in line with a higher purpose which will energise you instead. The health and general wellbeing benefits to doing this are substantial as well. Here are a few examples:

Decorating your home

Eco-friendly painting supplies mean you can literally breathe easy when decorating. As a minimum, look out for paints with no VOCs (Volatile Organic Compounds) which release the toxic smell. For maximum bang for your buck, use a company that takes its environmental and social responsibilities seriously. Little Greene, in the UK, touts several environmental credentials including local manufacturing, products and packaging made from recycled materials, and child-safe paint for children's toys on their website.[125] In the US, one option you can try is https://www.ecospaints.net/. They produce non-toxic paints that are made to order in the US, so no waste from products being made but not sold, and no overseas shipping involved.

Household supplies

Things like cleaning products, dishwasher tablets, and laundry detergent don't have to be toxic. To clean, I either use white vinegar or water with tea tree oil which is all that's needed for everyday jobs. It saves money and a lungful of toxic dust which, according to a 2018 study, is as bad for lungs as smoking 20 cigarettes a day.

"When you think of inhaling small particles from cleaning agents that are meant for cleaning the floor and not your lungs, maybe it is not so surprising after all."[126]

Indeed.

Personal hygiene

Try products such as shampoo, conditioner, and shower gel (or a bar of natural soap) made from natural ingredients. I've been using 'Faith in Nature' and refilling the bottles at the local health food shop. I make my own deodorant and moisturiser now. I used to spend a fortune on expensive brands of all this stuff. Sucked in by the advertisers' gimmicks that promised more youthful skin and/or radiance, they lured me in with promises of successful 'clinical trials' and fancy made-up scientific(ish) words. Of course, it's all BS. None of it made any real difference and probably did more harm than good.

Senior policy strategist for the non-profit Breast Cancer Fund, Nancy Buermeyer, says, "We don't know enough about (consumer) chemicals on any front and certainly not about how they impact women because we haven't spent the time or energy to look at it."[127]

Not only are we declaring ourselves lab rats by constant exposure to unnecessary chemicals, but I can safely vouch that the natural alternatives are by far the most effective products I've used. If you want to have a go at making your own, you'll find plenty of recipes online.

Think outside the gift box

Thinking about how we give gifts has made a considerable impact on my life. I get so much more pleasure out of it now. And I think the people who receive them do as well.

A few years ago, like a lot of us do, I was buying presents for Christmas and birthdays out of obligation as much as anything else. I would traipse around the shops and spend time searching online for something I thought the person would appreciate. Maybe some clothes, accessories, makeup, booze, or a gadget. I genuinely wanted to buy something they'd like and, under time pressure, trying to find that perfect thing was more stressful than fun.

Receiving gifts often wasn't much better. It's lovely to be given something, but no one knows your taste like you do. And, you're never normally sure if you like something *the first second* you lay eyes on it; so you have to have your 'I love it' face all lined up when you're opening it.

The problem definitely isn't with gift-giving. There's nothing better than randomly coming across something and immediately thinking of the person that would love it because it signifies something meaningful between you. The problem is how gift-giving has become commercialised. In other words, *forced*.

We've built into our culture that materialistic trinkets *must* be given to each other on specific occasions; the prime example being Christmas. We're made to feel that if we don't begin a New Year almost bankrupt and broken from our shopping sprees, then we're doing something wrong!

I have a friend to thank for initiating the change from a consumer Christmas to a conscious one.

A few years ago, she sent a message to everyone that might buy her a present saying she was giving the stress and expense of buying gifts for the sake of buying gifts for Christmas a miss. She explained that she was ditching the approach in favour of buying gifts for people when she sees something meaningful that they would really like. She encouraged everyone to do the same or donate money to a charity on her behalf instead of getting her a gift.

A few of us on a WhatsApp group got on-board straight away, and I adopted a similar approach with people I know. I ended up just doing a Secret Santa with immediate family and only buying presents for the kids. I've kept this up ever since which means no stressful rush around the shops trying to find the perfect gift only to end up with a bottle of wine and a cat calendar. And, I'm fairly hopeful that no one likes me any less than before. In a word – it was *liberating*. Not to mention how amazing it feels to have money going to charity compared to having another ornament on the shelf.

We saw in chapter 3 that regularly buying consumer products increases our carbon footprint by as much as 4 tonnes. Going crazy at Christmas forms a chunk of this total, so lighten your carbon load by thinking a little differently.

Choose (experiencing) life

When you do want to buy a gift for someone, there are two excellent reasons not to buy them a manufactured product.

One, studies have shown that experiences make us happier than material things. It's not hard to imagine your loved one getting a lot more pleasure from a dinner date with you, or a show, a massage, a bungee jump, or horseback ride – or whatever they may discover they're into – compared to a new jumper.

Two, by swapping manufactured goods for 'paying for a service', the average EU consumer would cut close to a hefty 12 percent of their household carbon footprint.

"Consumers have a huge environmental impact: we like to blame the government or industries for the Earth's problems, but what we buy makes a big difference."[128]

Toilet paper

Chances are, you probably already use recycled or sustainable toilet paper. If you don't, then make this swap ASAP because – according to National Geographic – toilet paper is responsible for wiping out 27,000 trees a day! It comes from forests around the world, including North and South America and Scandinavia. Some being hundreds of years old – such as those in Canada.

On top of swapping, you might also want to consider which company you buy from. You can buy a supermarket's own brand of recycled toilet paper, but if you want your consumer power to pack the biggest punch then buy from a company that has social goals as well as profit-making ones. Backing a company that only sells environmentally considerate products will encourage other businesses to adapt to the new culture we create of respecting the climate. Personally, I buy from Who Gives a Crap. They donate 50 percent of their profits to help build toilets in the world's most underserved regions.[129]

My final observation on toilets is – do we need TP at all? The Japanese have come up with what seems to be a far more civilised and less wasteful method. Japanese toilets use water and warm air to clean and dry you, and their clever design means they're water and energy-efficient. If your bathroom needs a refurb, then look into getting one. The cost varies, but for the less expensive models, you'd make the money back from toilet paper savings in the first couple of years.

Packaging

Having been trained in Lean/Agile, which is all about cutting out waste, single-use packaging has long been a bugbear of mine. And I know I'm not alone.

With food and a few home products, we can do our best to avoid stuff in packaging. We can use the local shops that offer refills and bring-your-own containers instead. But for pretty much everything else, packaging is hard to escape. This is one case where we really need businesses to take the initiative. Fortunately, it seems like there might be hope on the horizon.

TerraCycle – the organisation talked about in the Recycling chapter – have partnered with some of the biggest consumer product companies to offer a shopping system and reuse model they're calling 'Loop'. The idea is to reduce single-use packaging and instead offer a circular packaging model for branded products. Some of the brands currently involved in the scheme include Nestle, PepsiCo, Unilever, The Body Shop, Coca-Cola European Partners, and Danone.

Chief executive of TerraCycle, Tom Szaky, has said, "Through Loop, consumers can now responsibly consume products in specially-designed durable, reusable or fully recyclable packaging made from materials like alloys, glass and engineered plastics. When a consumer returns the packaging, it is refilled, or the content is reused or recycled through groundbreaking technology."

The products are either delivered to your home through the Loop website or a brand's own website. And the packaging is collected at the same time to be reused, or you can buy and drop off the packaging in store. In the UK, Tesco launched a trial for the scheme in 2020 for online shoppers.

Some of these brands are not known for their environmental commitments, to say the least, so I'm not suggesting you switch to them or buy from them if you don't normally. But if you do, and you're not ready to swap out the products you buy from them, yet, then at least we can get behind this initiative for the things we genuinely need. This will give the concept of reusable packaging the chance to flourish, and the opportunity will emerge to cut out wasteful packaging for good. Go to loopstore.com for more info and how to buy the products.

On your phone (or tablet or computer)

I've come across a couple of interesting online lifestyle choices that can make an impact on climate. These include our email, internet search, and streaming habits.

Streaming

I wasn't dedicated to a streaming service when I read the Greenpeace report from 2017 on which providers invest in renewable energy.[130] That has made it easy for me to use the companies that score the highest on the report and minimise my use of the others. (The report scores on renewable energy commitment and procurement, and energy efficiency and transparency.)

The companies that fared well were Google, Facebook, and Apple. That includes iTunes, YouTube, and WhatsApp. Those whose performance was poor included Netflix, Amazon, and Spotify.

That may have just made your heart sink if you use one of the latter. Streaming technology is now a big part of our lives, and this shows in the increasing contribution of the tech sector to GHG emissions. But, entertainment brings us a lot of enjoyment, which means the brand can tie us in with clever marketing. For music streaming, each brand would have us believe that they have the functionality we can't live without. I'm not a tech expert, but from what I can tell, they all offer pretty much the same functionality albeit with different programming. If you can bring yourself to do it – part ways with your music service provider until they demonstrate the necessary respect for our world. Then you'll be making your vote for climate safety count. TV is a little different as you might be tied into the shows you watch. But you can still think about how you watch TV in the future, and limit using the providers with the worst climate credentials.

Emails

The topic of emails is an interesting one, not just in relation to the environment but also our general wellbeing. I've read many times that the invention of email was supposed to make us more productive. But what's happened is that we are now expected to *always* be available and respond to each other much *faster* than before. In other words, we now work even harder.

Of course, changing our entire email culture is beyond the scope of this book, but – as with everything Lean/Agile – we can move towards our goals one step at a time. To reduce our emissions, as well as improve our sanity from constant email exchanges, we can cut out the unnecessary ones that are evidenced to cause harm to the climate.

An OVO Energy study revealed that Brits alone send over 64 million unnecessary emails every day, mostly just to respond to someone saying 'thanks'. This happens even when we're in talking distance of each other.

But the survey also revealed that most of us (71%) would be happy not to receive them if it saves on emissions. Most shocking is that the findings show that "If each adult sent one less email a day, Britain could reduce its carbon output by 16,433 tonnes – equal to more than 81,000 flights from London to Madrid."[131]

Plant trees while browsing

I mentioned that Google's sustainability credentials are pretty good. But we're looking to take part in the regenerative revolution – to replenish our planet with more than we take out. And also shape the businesses of the future to serve our wellbeing, not the financial god of growth. So, for us, there's a better option. It's a search engine called Ecosia. This not-for-profit uses revenue from ads to plant trees rather than line shareholder pockets.

"The Ecosia community has already planted millions of trees in Ethiopia, Brazil, Indonesia, Spain, as well as many other biodiversity hotspots... CO2 neutral is not enough. Thanks to our solar plant and the Ecosia forests around the world, each of your searches removes around 1 kg of CO2 from the atmosphere."[132]

There's a counter on the homepage that continually updates and tells you how many trees have been planted by users – it's a lot. The aim is a billion.

There's also a counter that tells you how many trees you are personally responsible for planting. I can't think of many other organisations that offer such a feel-good factor for such little effort and zero cost. The icing on the cake (for me at least) is that it's now available as an extension to Google's Chrome. Go to https://info.ecosia.org/about for more information.

2. Travel

We covered the most climate-damaging means of getting from A to B, and how to avoid unnecessary travel, in the chapter on Transport. But there will be times when we want to travel further afield to see different places and cultures so that we can enrich our lives. And when we do, we should bear in mind that tourism accounts for 8% of global GHG emissions according to a recent study.[133]

This is four times worse than previously thought. Fortunately, there are now more and more ways to travel, and places to stay, that keep the trip as environmentally-considerate as possible. As well as sustainable tourism, regenerative tourism that uses a circular model (i.e., replenishes destinations to leave them better than you found them) is emerging. And as always, our opportunity – as conscious consumers who want to shape a better world – is to support it.

Fortunately, this is the burgeoning era of the conscious consumer, so there are already several organisations out there who exist to better the world and our experience of travelling around it. Because their focus isn't fixated on profit, they're free to think about what other goals they should measure their success against. Like Tour Operators 'Bookdifferent' who started a "booking website where you can easily find green hotels and support a charity every time you book."

Or, 'Barefoot' – which is "...driven by a social and environmental ethos working in partnership with NGOs, co-operatives and social enterprises to provide more immersive cultural discoveries and enriching ecotourism adventures, away from the conventional and commercial tourist circuits."[134]

Other travel operators worth checking out include: responsibletravel.com and travelifecollection.com

To help you decide where to go, as well as where to stay, check out the Global Sustainable Tourism Council (GSTC) website.[135] They manage the GSTC criteria of "global standards for sustainable travel and tourism."

As well as booking through organisations like these, you can take your conscious consumer mindset away with you. Think about what you buy, how you travel, what you eat, and your energy use while you're there. Cut out anything unnecessary and think about how you spend your money, the same way the other chapters suggest you do at home.

There's one other thing you can consider as a conscious consumer that will have an impact on your carbon footprint. It applies to when you're at home as well as abroad. It's *localisation*, and it means considering the distance that what you eat and buy has travelled.

The reason we haven't covered it in the other chapters is because it's not the biggest priority on the footprint calculator we've been using. But it does have an impact and is well-deserving of discussion in this book. So, the third section of this chapter is dedicated to just that.

Other types of travel – you cruise you lose

I used to think that going on a cruise might be nice, but the more I've found out about them, the more I've realised how dirty and overindulgent they are. Cruise ships spew out some of the dirtiest emissions around, and it's not just impacting climate... our lungs are also taking a serious hit. "Carnival Corporation, the world's largest luxury cruise operator, emitted nearly ten times more sulphur oxide (SOX) around European coasts than did all 260 million European cars in 2017."[136]

To put this in perspective, in Southampton just 44 ships emitted around 27,000 tonnes of SOX compared with circa 3,000 tonnes from 260,000 cars!

The ships are hanging around various coastlines now, with nowhere to go as the Covid pandemic continues (it's common for diseases to spread rapidly when living in close conditions such as on these ships). And although they may look pretty, lit up at night, no one enjoys choking on their fumes when they fire up their engines. "Air pollution from international shipping accounts for around 50,000 premature deaths per year in Europe alone, at an annual cost to society of more than €58bn [$65bn]."[137]

As well as using dirty fuel – and lots of it to power the ships' propulsion systems – the industry is similar to flying in terms of emissions from the powering of all the on-board facilities. On top of that, passengers are encouraged to indulge themselves with food and drink laid on by the bucketload. Although some of the passengers may try, they can't possibly get through everything that's on offer. Which means waste. "A ship with 6,000 people on-board can generate around 2,100 tons of waste water, 24 tons of wet waste (food waste and bio sludge from waste water treatment plants) and 14 tons of dry waste per day (solid burnable waste, plastic, glass, tins and cans). All this waste altogether is enough to fill around 110 trucks."[138]

Based on all this, I wouldn't touch them with a bargepole, and I hope you feel the same now if you didn't already. However, if it's been your lifelong dream to go on a cruise, but you're put off by all this, then there may be a more sustainable option. Luxury French cruise line 'Ponant' operates a fleet of smaller ships and are committed to sustainable tourism goals. They have ships powered by Liquefied Natural Gas and electric batteries instead of dirty fuel. They also carry out studies to help ensure a minimum impact on the ecosystems their ships travel through.[139] [140]

Learn how to fly

If there's no other way than to fly, then – believe it or not – there are ways to do so (slightly) more sustainably.

Fly direct. Carpool on your way to the airport or get the coach or train. Fly from your local airport when possible.

You may want to research which airlines are taking sustainability more seriously before choosing which one to fly with. For example, KLM is actually encouraging people to fly less and uses biofuels on some routes.

Furthermore, KLM and All Nippon Airways are both in the top three of the Dow Jones Sustainability Index. This means that, relatively speaking, they are amongst the least polluting airlines.[141] [142] [143]

Extra resources

The Guardian has published a helpful guide to flight-free holidays.[144] Clearly, it is more relevant for trip planning if you're UK-based, but there's plenty of interesting information on how to be a green traveller regardless of where you're from.

You can find out your carbon footprint for specific journeys here: http://www.ecopassenger.org

3. Localisation

You may have heard about some of the crazy things that happen in our new, consumption-driven, globalised world. Apples being flown from the UK to Africa to be waxed and flown back again.[145] Or, similarly, fish being caught, frozen, shipped to China to be processed, and then shipped back again.[146]

This bonkers behaviour isn't just happening with food importing and exporting; the same craziness happens across industries across the world.[147] It seems profit doesn't just come before the planet, but before common sense.

The sad thing is that because we've grown accustomed to being able to get hold of anything from anywhere for the lowest price, we now expect it. We feel like we're missing out on something if we're not able to pick up exotic fruit all year round. I know this feeling well as I'm partial to avocados and the odd pineapple.

The thing is, with this mindset, we might be missing out by not shopping locally, and not just in terms of indulgent titbits. Economists and other thought leaders who are interested in sustainability are telling us that our wellbeing will hike up if we invest in our local economies.

Probably the most famous of these is economist Helena Norberg-Hodge. She believes that to make the fundamental change to the economy that is needed, we should shift it from global to local. This will have multiple benefits, including a reduction in CO2 emissions, energy consumption and waste, the restoration of biodiversity and cultural diversity, and the creation of meaningful, secure jobs for the entire global population. In turn, we get to increase our happiness by rebuilding the fabric of connection and community amongst ourselves, and with our local environments![148]

All that said, localisation isn't the first thing you should do to make maximum savings on your carbon footprint. The products and services we spend our money on (and how often) have a bigger impact according to footprintcalculator.org. This is why I've talked about cutting down on, and swapping, your current consumer choices in the other chapters. There is a 'but' here, however. Once you've at least begun putting these improvements in motion, localisation should be the next thing you consider.

The good thing is that by making some of the swaps suggested in the other chapters, you will have already been replacing some of your corporate purchases with local ones. Rejuvenating clothes using a local clothes alteration service, upcycling your furniture rather than buying from Ikea, and staycations are all examples. Now, let's look at how we can take it a step further.

Local food

The footprint calculator we're using gives a carbon footprint saving of up to half a tonne if you swap all your non-local, packaged foods for the local, unpacked kind. A decent chunk of carbon to chip away at. And, as what you'll find in your local food retailers can also prove far healthier than what's on the supermarket shelves, the wellbeing benefits are considerable too.

Probably the biggest movement towards localisation is with food. It may be that you're already a regular down-the-farmers-market type of person, or that you get most of your food from local farm box deliveries. Or, it may be that, like me, you find the convenience grip of supermarket shopping tricky to wriggle out of. It took me lots of conversations with friends and an understanding of footprint savings (and other benefits involved) before I started to make my shopping habits more local. And, as always, the continuous improvement approach is what has worked to make changes actually happen.

I didn't suddenly put down shopping trolleys at supermarkets for reusable containers at the local wholefood store. And I'm still easing my way away from them now by trying one health food shop and veg box option at a time. And I'm making lots of satisfying discoveries on the way. Like...

- How much fresher the nuts taste (and no doubt are), and how good it feels that they come packaging-free or in compostable packs.
- The range of organic pasta substitutes that are made out of healthy lentils and legumes instead – a really easy way to make tasty vegan dishes packed with protein.
- How much better it feels to find a local wholefood shop that sells non-packed frozen fruit (a staple in my diet) instead of buying the non-recyclable packed stuff from a big supermarket.
- And discovering the extensive range of UK-produced organic dark chocolate on offer!

Even if you're already pretty good at avoiding supermarkets, you might want to refresh your knowledge of the alternatives in your area. You might find a new local wholefood shop or farm delivery option has opened up near you. Do a bit of digging online and spark up conversations with friends about where they're shopping these days. I normally find that opens my eyes to more and more options out there; it makes the urge and the need to shop at multinationals less and less. And another added plus is that local food shops usually sell more than just food. The other products they sell are often local, and more ethical and sustainable as well.

Local clothes

The carbon footprint calculator we're using doesn't give a carbon impact from buying your clothes from abroad. Probably because it's so commonplace, it seems hard to avoid it. But there is something you can do.

Find a website or blog in your country or region that gives you a list of sustainable clothing produced locally. In the UK, frombritainwithlove.com gives a list of 15 fashion brands that actually make their clothing in the UK (you can get other local items from this site, too). Sorry boys, they seem to only be for the ladies.

The next best thing is to buy from a UK brand, or a brand from whichever region you're based in. This will give you a lot more choice and a quick internet search will bring up a multitude of lists to choose from in magazine, newspaper, and blog articles. Okay, they may not make the clothes in your country or region, but they will distribute from there. Assuming we're not buying from the high street due to the lack of sustainable options, this will cut out the extra carbon emissions from shipping your sustainably-sourced clothes to you from abroad.

Feel the wellbeing benefit

Your experience may well be different but, personally, I haven't made any best friends by shopping locally – yet. Regardless of that, shopping locally definitely feels like you're part of something bigger. By supporting your local economy and small businesses – especially when they're offering sustainable and ethical products and services – you are building something better. A future where there's less need to work for a big corporation where we feel like a number. We will all have more opportunities to start businesses that provide something truly valuable to our local communities, because we know it will be supported by like-minded conscious consumers, such as ourselves.

Let's keep getting better and better

This chapter has only covered a handful of the myriad ways we can make little improvements that add up to big carbon savings.

"Big problems are rarely solved with commensurately big solutions, instead they are most often solved by a sequence of small solutions."[149]

Although I've reached the 3-tonne goal, I live by the continuous improvement mentality because of the value it adds to my life. I don't have to make changes that I find daunting, but with each small thing I do on top of the last, I feel that I'm contributing towards personal wellbeing today, and a safer, more prosperous collective future.

In Lean terms, the principle of continuous improvement is also known as striving for perfection. Striving is the best we can ever do – we'll never be perfect 'eco-warrior angels', of course. I think it's worth mentioning that point – considering all the negative comments that get bandied around in the media (social and otherwise). It's definitely on the primitive side of our nature to point fingers at people who are taking action on environmental issues but still use plastic products (for example), or drive, or eat meat. We're all going to be guilty of some environmental damage.

Instead of having a jab at someone, we should be capable of realising that it's unhelpful. If we want to aim higher, we can choose to support each other when making an effort and take inspiration from the actions of others.

"Do the best you can until you know better. Then when you know better, do better." Writer and civil rights activist Maya Angelou.

A final thought on a climate-considerate lifestyle

Thanks to the help from books and recordings by psychologists and spiritual teachers, I've been able to practise reflecting on negative thoughts. And I've asked myself who I really want to be.

This has been invaluable when making lifestyle choices, no matter how small, that affect the environment. If you're reading this, I'm going to assume that you share at least some of my aspirational identity that I continually work towards. So, I've laid it out below. I've included it, as I hope that reading and reminding yourself of it will help you make positive decisions when faced with the environmental challenges that everyday life brings.

I want to take action and some level of responsibility for the state of the world around me. I have an awareness that the consumption-driven system has sucked me in further than I'd like, and it doesn't lead to a contented life.

I've got a niggling feeling that I'd be better off going against the grain of a lifestyle that revolves around working harder and harder so I can buy more and more and bigger and more expensive things.

I think a better way would be to connect and spend more time doing what's truly fulfilling, like educating myself, or outdoor adventures with family, for example, and helping and guiding other people to live more fulfilling lives too.

I know I'll be happier and healthier if I can act on my inbuilt respect for the natural world, the part of me that's moved when watching documentaries like 'Planet Earth' in total awe. The part of me that knows my natural environment gives me a sense of peace. The part of me that feels a little bit guilty every time my food goes to waste, or when I buy something without thinking. I remember how amazing it feels to live according to my values.

I know that our beautiful home needs saving; it's in all of our hands, and I want to be a part of the movement that creates the global mindset change to turn it all around. If I can reach the goals and live like this, I'll be able to look younger loved ones in the eye and say I played my part when it really mattered. I'll also feel a sense of peace that only comes from transcending personal interests by doing something for the greater good.

Key takeaways from this chapter

- There are tonnes of environmental suggestions out there already, but unless we implement them with some idea of the impact they're having, we could be chasing our tails.
- Now that you understand the primary areas of your lifestyle that are causing damage, you can do some of the smaller stuff that adds up to a significant impact and makes good sense.
- Being a conscious consumer will have an impact on all industries. Once you've cut out unnecessary purchases, consider the eco-friendly alternatives before you buy.

- There are lots of healthier options out there which are better for the planet. From the chemicals we use in our homes and on our bodies to the streaming services and web browsers we use.
- When planning trips, use your climate-conscious mindset. Choose an eco-conscious tour operator, and take your good habits of what you eat, buy, and do at home, away with you.
- Start living local. On top of the other lifestyle changes in this book, localisation can have a hugely positive impact on society as well as climate.
- If you're struggling to make some of the changes suggested, it might help to remind yourself of the values you hold dear and the impact you want to have on the world. You might find the aspirational identity I provided helpful, or you might want to write one of your own.

9 | Building a Better Future and a Better World

Throughout this book, alongside the practical suggestions, we've talked about how an understanding of the root cause of climate change – the CC22 – can give us the awareness we need to make conscious consumer choices. And the Consumer-Led Movement can provide the platform to put *collective* improvements in place. This can reduce our carbon footprint and shape businesses and government policy for a better future.

This chapter further explores that vision for a better world and what is currently stopping us from getting there. Are there really viable alternatives for how we operate today? Or, is anything other than consumer capitalism and the neoliberal 'free market' agenda nothing more than a pipe dream?

In chapter 1, we talked about the ethos of capitalism as being about looking after ourselves first, and everyone else as an afterthought. And how today's capitalism measures societal success against the sole goal of 'growth in the economy'. These traits of the system bring out the worst traits in us and keep us stuck in the mindset of the primal brain. By comparison, we need to access the higher brain to be able to think up something better.

So that we can build a vision of the kind of future we're aiming for, we shall break this down further for, as Steve Jobs said, "If you define the problem correctly, you almost have the solution."

A closer look at the system

"Growth is what the capitalism system counts, needs and does." Giorgos Kallis

In a business setting, it's commonly understood that whatever measures are used to judge the success of employees will impact their behaviour. If, for example, the most important thing is for a team to work quickly, then you measure their speed. But, you had better realise that if that's *all you measure,* then speed will come at the sacrifice of performance in other areas. Ever spoken to someone at a call centre and it seems like they're more interested in getting off the phone than resolving your issue? You can bet your bottom dollar that they're being targeted on how quickly they complete the call, not whether you're satisfied with the outcome.

You may have found that this was a more common experience *in the past* than it is now. That's because many companies now realise that they need to measure customer experience, as well as speed, to resolve customer issues and keep our business. It sounds obvious. You've probably questioned it yourself after experiencing one of these calls previously. How on earth could businesses, full of intelligent people, not realise this? But the way in which *society as a whole* is measured drives all our behaviour – individuals and businesses.

Society measures the growth of GDP (Gross Domestic Product), i.e., the monetary value of goods and services produced by a country. As individuals, this means that many of us sacrifice family time and nurturing relationships to work harder and more, to make as much money as possible. Because that's how society judges success.

And businesses are measured on increases in profit. The focus on profit at the top, (e.g., with shareholders), filters down to the individual teams. This means measuring the true effectiveness of teams often loses out to measures like speed; do things quickly and save money to maximise profit. At least that is the idea.

In the call centre example, what they're not taking into account is that no matter how quickly the call finishes, if the issue isn't dealt with appropriately the first time, then the customer will have to call back. And possibly be annoyed enough to take their business elsewhere. And all efficiency gains are lost.

By not measuring other goals – in this case, whether the customer is happy that their problem is solved – we're missing the point. In society, there are more important things to measure than pure economic growth because 'money' and 'things' don't equal happiness!

In turn, economic growth, as we have seen, can drive the wrong behaviours because we'll do anything to maximise it, even at the sacrifice of what's really important. At a societal level, the health of the planet and our overall wellbeing is sacrificed in the name of perceived wealth (keep reading to see how the wealth of the masses has not increased since the growth goal became entrenched under neoliberalism). And as we don't have other measures in place to counteract the drawbacks of maximising growth, the damage that it causes is plain to see.

The damage is being done

We saw in chapter 1 why GDP is insufficient as a measure of success for society. Even its creator, Simon Kuznets, cautioned against using GDP growth as a proxy for economic or social wellbeing back in the 1930s.

"It's to our peril that governments' preoccupation with growth leads to other very important concerns – socio-economic ones and environmental ones being put on the backburner as they pursue growth at all costs."[150]

To demonstrate just how destructive the path we're on is, we only need to look at the facts:

- Capitalism has been around for hundreds of years and has always been a game of winners and losers, with a disparity between rich and poor.
- Measuring growth in GDP came along after the Second World War. It was only ever supposed to be *one* indicator of monetary wealth, not used as a means to show that all is well in society as it is now.
- Around the same time, consumerism went to a new level, propelled by the initiation of mass production and widespread propaganda campaigns designed to influence our buying decisions.
- The modern form of neoliberal capitalism began around 1980, when neoliberal policy reforms encouraged the deregulation of business, not least by the UK and US governments under Margaret Thatcher and Ronald Reagan.

Adding this ideology into the mix created the perfect storm. Since then, businesses have been free to pursue economic growth and profit maximisation at any cost, by peddling consumerism to the masses.

The damage this has caused is clear. Global CO2 emissions rose 60 percent between 1990 and 2013. That's *just* 23 years. This increase is attributed to 0.8°C of warming – the majority of the rise in mean global temperature since pre-industrial levels... 260 years ago.

Are billionaires to blame?

Neoliberals have convinced us that the *market*, the arena in which commercial dealings are conducted, is an all-knowing, self-regulating entity that should be left to its own devices. This will ensure a healthy economy. In reality, the market – as it stands – is only a concept thought-up by humans who have all kinds of subtexts, wants, desires and needs. The 'market' is subsequently always being governed by those who have the most influence over it.

Those who benefited the most since the neoliberal era began, some of whom try to convince everyone else that wealth will trickle down, are the 0.1 percent richest people. They captured as much wealth as half of the global adult population between 1980 and 2016.

For those who sit between the global bottom 50 percent and the top one percent – AKA, you and me – income growth since 1980 has been sluggish or non-existent.[151]

In fact, wage growth is the lowest it has been since Victorian times. And debt to GDP is at record highs.[152]

It's my thought that the richest one percent of people are the biggest victims of the Capitalism Catch-22. They are fully aware that the system expects financial growth, and are subservient to this goal above all else. This has them trapped in a cycle of succumbing to their lesser natures. However, I am not expecting anyone to pick up a violin.

Of course, some people with a lot of money give large amounts to charitable causes and spend time doing benevolent acts. I'm not suggesting that people with the most money are bad people. I'm suggesting that the system has all of us indulging the greedy part of our nature more often than the 'human' (higher thinking) part that values wellbeing for all. Hence, any charitable and benevolent acts we pursue as a society are not enough to close the gargantuan disparity between rich and poor. Or, make up for all the social injustices that brings. Nor are they enough to ensure we live within ecological limits.

Government failure

To avert climate change through government action, governments would need to ramp-up environmental policies that reduce consumption and limit growth, which also interferes with 'the market'. Sadly, anyone expecting this to happen – as things stand – are largely barking up the wrong tree.

As Noam Chomsky[153] puts it, "In the US, there is basically one party – the business party. It has two factions, called Democrats and Republicans, which are somewhat different but carry out variations on the same policies. By and large, I am opposed to those policies. As is most of the population." Perhaps governments would feel more pressure to ramp up policies that combat climate change (rather than serve the interests of big business) if they thought they would lose votes. But they know that we are all mentally invested in this system and are scared at the thought of changing it.

- That's why incumbent parties are confident that they will get voted in by promising economic growth above all else. And creating fear that other parties will not do that as effectively.
- That's why they don't pay attention to a few people shouting that the world is on fire.

What about us?

As well as tackling the system, we also need to tackle the other part – our less desirable traits. After all, these did come first.

As the adage goes – the more we have, the more we want to have. Not just in terms of 'things', but influence as well. It's the age-old story of wanting to rise to the top of the troop that our primate ancestors would've done, and our primate cousins still do.

Now, we have evolved a higher thinking brain, but we still have our primal brain, and this is the strongest part. Psychologist and author, Professor Steve Peters, often talks about the part of our brain that controls our emotions and says it's like having a chimp who actually lives inside our head! This chimp overpowers our rational thoughts. So emotions, including greediness and competitiveness, often win out.

But our emotional behaviour is not always *who we want to be* when we use our higher thinking brain to analyse it. Peters suggests that when we're using our higher brain capacity, we're connecting with our true selves. This "is the real person, it is you; rational, compassionate and humane, and is the Human within."[154]

As our animalistic instincts have been compounded by the system we're living in; the part of us that 'wants more' thrives, while the higher thinking brain has less opportunity to flourish. No wonder the culture of wanting more is completely ingrained into our society. But striving to obtain more and more wealth will never completely fulfil us. Logically, it seems that the opposite is true – striving for more is likely to bring us the continued desire for more, which will leave us always wanting!

But if we ditch capitalism and the growth goal, won't we be ditching prosperity and innovation?

Innovation and the profit-making incentive

People often argue that capitalism is a catalyst for innovation. In some ways, this seems to be true. Capitalism has been the setting for some of man's incredible achievements in the last couple of centuries. Mind-blowing ingenuity has enabled civilisation to develop at lightning speed thanks to technological and medical advances and new ways of thinking. That said, it's worth questioning whether the profit motive of capitalism is entirely responsible. Or, whether we've achieved some of these things in spite of the system.

Once we have our basic needs catered for, humans have an inbuilt craving to want to progress, develop, and be creative, but these things come from the higher brain. Many things we've achieved have come through collaboration and knowledge sharing and building on each other's ideas, which the competitive spirit of capitalism, (i.e., having to keep knowledge within the company to increase market share and maintain advantage), discourages.

We just need to look at the many advances that have occurred under an academic model rather than within profit-driven businesses. The initial Google algorithm of webpage ranking was developed at Stanford University. And university research is responsible for hundreds of innovations from seat belts to solar power to chemotherapy drugs.

UNIVERSITY RESEARCH

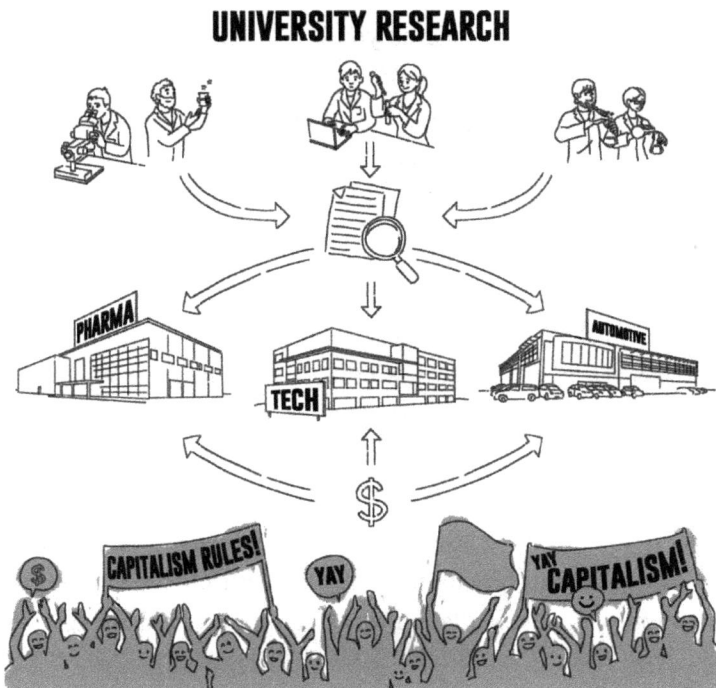

Let's not forget good old Wikipedia and other digital, cultural and intellectual commons which rely on the generosity and expertise supplied by the public, for the public.

Capitalism measured by economic growth may have been the backdrop for recent innovations, but if we believe that human ingenuity wouldn't and couldn't prevail under alternative systems, we're doing ourselves a disservice. Would we have advanced society in terms of material achievements as quickly under a different system? Maybe not. But achievements concerning the not-so-tangible elements to fulfilment have been sacrificed in its name. Progression relating to the greater good of humanity – such as finding a sustainable way to live – has been pushed to the side-lines. The profit motive that drives innovation in our capitalist system will always favour the type of creativity that will bring in the most cash, not the type that benefits the overall wellbeing of society.

Capitalism = prosperity?

There are many skills vital to our wellbeing which capitalism doesn't value. The system is only fair as long as you're playing the capitalism game. If you're prepared to maximise profit by supplying products people want (though often not what they need) – even if this means keeping wages, working conditions and environmental concerns low – then you get the 'glory'. If what you have to offer the world is your listening skills, you're not seen as having any value at all.

If you work hard at social care, teaching, or healthcare, or any number of vital roles in society, your financial reward is minimal because you don't have business acumen. If you're a homemaker, an environmentalist, or a poet, (unless you're one of the very lucky few), then forget it. The Covid-19 pandemic happening at the time of writing this segment is showing us just how important these job roles are in society, which capitalism completely overlooks. Capitalists who think that the system is fair, because everyone has an opportunity to make themselves rich, are *missing the point*. There are more valuable skills in the world than making money. And, in a truly prosperous economy, we can all live comfortably by offering them to the world.

Thank you and goodbye

Whether you believe that modern capitalism is being given enough credit for our societal achievements or not, the fact remains that the culture of wanting more has now got us – as a species – into serious trouble. We may owe capitalism and economic growth for our current standard of living in developed nations (and for that, we can be grateful), and the system has served us on some levels in the past, but it cannot serve us to the same level in the future.

Historically, it may have seemed to be working; at least, on the surface. There seemed to be enough resources to provide ever-increasing goods and services. And the output of these wasn't taking the planet beyond its boundaries. Environmental damage was being done, *but the ecosystems and atmosphere could cope.* And wanting more was a genuine need for a bigger proportion of the population who had a low standard of living.

But things have changed. The world is now different.

I am not suggesting that economic growth (albeit a 'greener' kind) should necessarily be phased out for those nations yet to experience it, but I am making the point that – for the rest of us – things are different. We're running out of natural resources that the world has to offer, and planetary boundaries have already been breached, especially in relation to climate change. We've grown so much that we're banging our heads on the planetary ceiling.

Now we need to move towards something different that will serve both us, and our natural environment, into the future. If we've come this far in the blink of an eye, using a system that limits us, imagine what we can achieve when we adapt to a system that facilitates more use of our higher thinking brain. It's my belief that many of us are ready to find out.

Shrinking down to size - growth alternatives

There are alternatives to a growth-economy which we'll touch on here. Luckily for us, lots of brainboxes have been worrying about how we can transition from what we have now, to something that considers societal and environmental wellbeing as fundamental goals. It's beyond the scope of this book to explore them in detail and, in reality, whatever we adapt to will end up looking different to what we imagine. It may be a combination of the below, or it may transform into something different due to real-world pressures. But it's still useful to have a vision for the future in our heads as we drive the Consumer-Led Movement forward. Because having an aspiration of where we're going means we know the steps we're taking are putting us on the right path.

Green Growth

The World Bank and the Organisation for Economic Co-operation and Development (OECD) promote a strategy of Green Growth, where growth is based on monetary spend rather than resources. As we spend our money on consumer goods that create a lot of emissions, Green Growth proposes that we can decouple the link between growth and carbon emissions. We can do this by spending more money on things that are less resource-intensive and use resources in a more sustainable way. Green Growth also relies on technological advances to ensure that depleted natural capital can be substituted. This is because, if we continue to grow the economy, existing technology is not capable of producing goods and services in a way that can reduce carbon emissions to the level required to avoid catastrophic climate change. Policies to encourage Green Growth could include: research and development into technological innovation, investing in low carbon infrastructures, and taxing products with high carbon emissions.

De-growth

This movement promotes downscaling consumption and increasing societal wellbeing. And an economy that 'sustains the natural basis of life'. In this society, humanity would understand itself as part of ecological systems and there would be a complete rethink of economic growth as a success measure. Instead, wellbeing would be increased by cooperative means, e.g., creating "open, connected and localised communities," and through "gift-giving". The proponents argue that examples are already being lived out in the forms of the voluntary sector, cooperative ventures, Wikipedia, social movements, and many more.[155]

Post-growth

The post-growth approach also advocates using wellbeing measures other than growth to measure the success of the economy. The difference to de-growth is that the approach suggests it doesn't have the definitive answer to the limitations of the growth problem but aims to develop and connect existing systems, technologies, and initiatives to evolve solutions which are appropriate to their specific time and place. These initiatives include Not-For-Profit organisations, circular industries to replace our 'throw-away' culture, and Credit Unions (member-owned financial cooperatives). It also takes 'planetary boundary' measures into account. The Post Growth Institute website states, "One-planet living acknowledges that we can, and must, mould our economies to fit within the limits imposed by our physical environment."[156]

Resource-based economy

There are also proponents of a completely different type of system which employs a 'resource-based economy'. This is a holistic approach where there would be no monetary exchange at all. Instead, all the world's resources would be the common heritage of everyone. According to the Venus Project website, it provides a "global socio-economic system that utilizes the most current technological and scientific advances to provide the highest possible living standard for all people on Earth."[157]

A note on 'Green Growth'

The research that's been done on modelling future 'green' growth scenarios shows us that avoiding catastrophic climate change whilst using traditional measures aren't possible. Growing the economy and at the same time reducing carbon emissions just won't happen fast enough, if ever. "Emissions reductions in line with 1.5°C are not empirically feasible except in a de-growth scenario.

"Models that do project Green Growth within the constraints of the Paris Agreement rely heavily on negative emissions technologies that are either unproven or dangerous at scale."[158]

In other words, if growth – green or otherwise – continues as predicted, the only way we can stay within the tolerable limit of warming to keep civilisation bearing a resemblance to today, is to rely on technologies that are (currently) highly speculative.

Aside from the above, we have also explored that the traditional growth measure promotes a culture of wanting more, which doesn't do us any favours. Ultimately, we have to stop focussing on it for a positive future.

The scary thing is that the Intergovernmental Panel on Climate Change (IPCC) scenarios for keeping global warming below the 1.5 to 2° Centigrade guideline, assume the economy will continue to grow. Even if we have complete confidence that governments around the world will act on their Paris Agreement pledges, in a growth economy it won't be enough to avoid catastrophic climate change.

This is yet another reason why we need to take our future into our own hands…

Building a bright future

Less growth – more jobs, more fun

One way to avoid unemployment in a non-growth economy is to share work between more people, which effectively means working fewer hours and having more time for leisure activities. The first time I heard this idea of working less, for less money, was on a Ted Talk titled "Plan B - is there an alternative to economic growth?"

At the time, even though I'd been toying with ideas about how society could be less greedy, it took some real soul-searching for this idea to seem anything less than ridiculous to me.

Even the very idea of less work for less money made me realise just how ingrained the need to increase our personal wealth is to us. Candidly, I found the prospect of sacrificing material things in return for more free time crazy!

I argued with myself that you need more money to be able to enjoy that free time. But then I asked myself – *what* is making me think that? Is it the truth that spending time with family or friends – even doing the most basic of things like reading and going for walks – isn't fulfilling, and we should be working as many hours as possible to plan extravagant holidays and saving for a bigger home instead? Or is this thinking the result of years of manipulation from cultural bias, supported by advertising and the media?

Hopefully, people are more open to the notion of working less and spending more time with family since the Covid-19 lockdown. It certainly sparks the kind of different thinking that we need to get us going on the path to a sustainable economy.[159]

In "Prosperity Without Growth", economist and author Tim Jackson recommends several ways to avoid unemployment in a non-growth economy. One is that rather than encouraging consumer spending to stimulate the economy, the government can invest in green industry and public goods. We've seen in examples throughout this book that investment in the 'green revolution' creates jobs. And to back up this point again – the required increase in renewable energy to meet the climate challenge is said to create more jobs than currently exist in the coal, oil, and gas industries.

"E2's recent Clean Jobs America report found nearly 3.3 million Americans working in clean energy – outnumbering fossil fuel workers by 3-to-1." [160] According to the Forbes article, America's two fastest-growing jobs in the next few years will be solar installer and wind technician.

Jackson also proposes that we could work in local or community-based social enterprises such as: local farmers markets, community health and fitness, music and drama, gardening, yoga, crafts, and local repair and maintenance services, amongst others. These types of activities are what many of us *hope* to partake in one day. But the way the world currently works, we have to put any move off until after retirement. By which point you've given the best years of your life over to the goal of maximising profit for shareholders.

Instead of that, we can have the opportunity to embrace more of the life we want now… as well as take care of the climate for a comfortable future for our younger loved ones and ourselves. This isn't a utopian fantasy; it's a viable option if we *choose* it, or at least *choose to move towards it* and see how it works in reality.

Out with the old

"The inability to imagine a world in which things are different is evidence only of a poor imagination, not of the impossibility of change." Rutger Bregman

Our aim is to drive collective change with the Consumer-Led Movement by reducing what we buy, and choosing what we do spend our money on wisely.

This is our plan of action.

And our vision is a world where we have more time to connect and enjoy nature because success is measured by societal and planetary wellbeing instead of growth. Now we need to get going. We have no more time for detailed debating and planning, or we're in serious danger of arguing ourselves to death. We must get moving now in the right direction.

If we start living by our plan, governments will get a clear message that we don't worship the god of growth anymore. If we show our leaders that we no longer want to create more and more wealth only to grow the bank accounts of the richest few, then – all of a sudden – it will be in their interests to respond to us rather than serve them.

Industries that are causing the biggest issues will have to adapt or die. This is true already and happens all the time; businesses that cannot respond to changing demands don't last. In response to the climate disaster we're facing, this will happen on a much bigger scale. But as some industries fall, new green industries, not-for-profits, and small, local businesses can take their place. And when government responds by investing in them and in us

– as we'll be the ones creating them – we should be able to avoid the negative impacts.

This isn't about bashing big business for bashing businesses sake; some of them care about the wellbeing of their employees and their impact on environment. All big businesses don't need to disappear, but they will need a steer in the right direction from us. We'll need to demonstrate that we support their sustainable practices, services, and products, until they reach a point of being regenerative (replenishing the environment by more than they extract) rather than destructive.

We know, however, that there are some businesses, and shareholders, who are quite happy to continue maximising profit at the cost of everything else. So, our vote against them, using our hard-earned cash, needs to send our message clearly: we will make conscious buying choices and re-distribute some of the world's wealth amongst ourselves. When our money stops lining billionaires' pockets, they'll no longer have the power to make the economy work for them rather than society and the planet.

Whenever I hear about the redistribution of wealth, it's as if the only way this is possible is through politics. It's as if that wealth hasn't come from us but somehow magically appeared in the billionaires' bank accounts without any conscious transference from our wallets. We could say there's an element of truth to that as we often spend our money mindlessly, but it's time to wake up and smell the wildfire. We are the ones who are channelling wealth to billionaires, and we are the ones who can choose to redistribute it, starting right now.

For successful civil resistance campaigns, "There are other security elites, economic and business elites, state media. There are lots of different pillars that support the status quo, and if they can be disrupted or coerced into non-cooperation, then that's a decisive factor."[161]

We can start to disrupt the status quo and shift culture, and with a much smaller percentage of the population than is needed to win an election. This shift can move society away from judging success through financial growth, by quite literally no longer buying into it. Then government policy is likely to follow. This should mean that we'll see a change in their investment strategies and environmental policies to support a new non-growth economy. An economy that we have *voted for with our money and our feet,* because we're ready to shed our label as 'consumers' and see what more we can be. Instead of being dazed and confused passengers, careering off a cliff in the back of the billionaires' rides, we will be the ones driving the economy – bit by bit – in the direction we need it to go for a bright future for us all.

Stop waiting, start creating

Change is scary for a reason. Big sudden changes can have disastrous results, so our brains are often doing us a service by being hard-wired to avoid them.

For example, some people believe that the answer to the current climate and social crisis is revolution. Aside from the fact that revolutions are usually very bloody and unpleasant affairs, initiating one is unlikely to do us any favours. We'll be just as likely to end up with something worse on the other side without a vision for the future. Or, without a step-by-step approach to improvement that allows us to adapt the solution to fit our needs as we go.

The thought of leaping to a new system is rightfully daunting. The idea of giving your money to organisations that put surplus money towards societal and environmental wellbeing – instead of focussing on private profits – isn't.

The move towards a regenerative rather than a destructive future has already begun thanks to our awakening consciousness about the plight of the living world and our place in it. This is without any collaborative effort from consumers. It's the way that culture is progressing anyway; we just need to make it happen quicker. And with a bit of direction and participation, we absolutely can! The more we support climate-considerate organisations, the more there will be as they rise to meet our demand and fill in the gaps left by the outdated business models. The ones that put profit first and foremost become obsolete.

Instead, we'll give our money, time, and awareness to organisations like: Not-for-profits, co-ops, charities, small-scale environmentally-focused entrepreneurs, local businesses, and community initiatives. Excess wealth created by these types of organisations goes towards innovation, increasing employee wages, and social and environmental goals, instead of into shareholders' pockets.

It's worth pointing out that by breaking free from a primitive mindset and giving our money to innovative, caring, and compassionate entrepreneurs, we create more opportunities to become one ourselves. "When you look at what's really going on in society, we've built global systems to keep people on the treadmill of being in the monkey brain," writes Daniel Priestly in his book "Entrepreneur Revolution". He suggests that tapping into our higher brain enables us to do something with our lives that we're passionate about. And that at this point in history – in the new age of accessible technology and information sharing – there are more opportunities to build something up than ever before.

When it comes to our vision of the future, we don't need to have all the answers; we just need to be open to new possibilities.

Because the trappings of modern society have emerged from our overpowering primal nature, they are indeed seductive. And the goal we base the success of society on is purely materialistic because these parts of our nature have been left unchecked. Knowing this, it's not too hard to understand *why* we've been unable to pull ourselves away. Understanding the root causes of issues is what gives us the power we need to break free. The modern understanding of psychology, as well as problem-solving frameworks, back up the notion that in order to solve our problems – we first need to understand them and their causes.

Getting to this point means we can set about improving our lifestyles. Step-by-step we can begin living differently. I know this to be true from personal experience. I've applied this approach to make easy refinements to my lifestyle that help prevent climate change. It's so much easier not feel the need to buy new clothes, gadgets, and general stuff regularly when you realise it's not what's really important to you. And you clearly see that advertisers are preying on your emotions so they can coerce you into consuming.

It's also completely liberating. I have found that not feeling the urge to buy new things, and spending money thoughtfully, represent a much happier way to live. Consumer lifestyle choices that are considerate to climate change can improve our wellbeing now. From avoiding Ikea and refurbishing furniture with a story, to backing brands that put people and climate first, and donating to charity instead of receiving birthday gifts – the uplift in satisfaction I've felt from doing all these things and many others is significant. This book has given you the suggestions and techniques to get you going on this path. Now it's up to you.

Hope for the future

As mentioned, I'm writing some of this book during the lockdown period due to COVID-19. For me, personally, it has been both scary and hopeful in terms of what it could mean for our future. Scary because we've witnessed how quickly people go into survival mode at the slightest hint of food scarcity (or a lack of toilet paper!), and how desperately and brutally people suffer and sadly die when our health services are overrun.

We're usually the ones sitting pretty in the developed parts of the world, watching the news and commenting how awful it must be for people struggling in war-torn countries or places with food shortages. This situation, although we are still the lucky ones comparatively, has given us a small taste of the future that's being predicted as a permanent and deteriorating state of a world where **climate change has run riot.**

This terrifying and suddenly very real flavour for the future is still in my mind. But, more recently, it has been superseded by a sense of hope for the future that I haven't felt since before I began researching the impacts of climate change, several years ago.

From the midst of all the current tragedy, I only hope that it's not all for nothing. I hope that we can all begin to see that we have an opportunity to think differently about our future so that it is bright and full of promise, instead of loss and suffering.

We've proven to ourselves that we don't need to buy stuff all the time; clearly, other things in life are more important. We can use this time to show our governments that we support something other than consumerism for our future. So, grab this opportunity by supporting all types of local, environmentally-friendly initiatives, community offerings, and businesses that arise. As time passes, you can choose to spend your money on experiences that will enrich your life and give you lasting memories.

Individuals and business owners alike are realising the benefits of working from home, spending time with family, breathing pollution-free air, taking bike rides and walks in open space, and spending less time in the shops. It's given me a glimmer of what it *could* be like in a society where people work less and enjoy non-materialistic pastimes more. And it doesn't just seem feasible, but preferable.

With each product *we choose not to buy,* or *choose to buy thoughtfully,* we'll be telling the world how we want it to be, and it *will* respond. We have an unprecedented opportunity to think differently and take our future into our own hands. The age of the conscious consumer is dawning, and we can speed up its actualisation by acting together. We have the impetus and now, through the Consumer-Led Movement, we have the means. Be part of the movement and, by your actions, create a future with hope for us all!

Go to consumerledmovement.com now to register your participation and keep updated on progress.

Key takeaways from this chapter

- The way we act and think, both as individuals and within organisations, is largely driven by the system we're living within and how it assesses success.

- Our system relies on, and measures, financial growth; hence, we put this above other equal and more important goals.

- This is exacerbated by the neoliberal, consumer capitalism of today. This means businesses are free to push products onto consumers that often don't meet our fundamental needs and take the planet beyond its limits.

- The richest people in society are the ones who benefit from the system as it stands. The rest of us are busy working to grow the economy for their financial benefit.

- This doesn't mean all rich people are bad; it means the CC22 is probably having the biggest effect on how they behave compared with anyone else.

- Governments are also operating within the limits of the current system, so they too aren't able to do what's required in the timescales needed to avert climate catastrophe.

- We, as individuals, have the most power and the most to gain from breaking free of the CC22 mindset. As soon as we embrace how the current system has major flaws, and there is another way to live, we can begin to build a society that works for us and the planet.

- The power is ours, and the time is now. Go to Consumerledmovement.com and add your name.

Appendix | The Lean/Agile Approach

How to be successful

There are tonnes of positive things happening to prevent the end-of-civilisation-as-we-know-it scenario that's being predicted because of the rapidly changing climate. I wanted to write this book because some of them, although well-meaning, are not actually helping. I wanted to show people what *will* help, and by *how much*, and how this problem is actually solvable if we focus on the right things and implement them successfully.

Thanks to my career – that has focussed on complex problem solving and business improvement – I have become very familiar with Lean and Agile techniques. When it comes to making lifestyle refinements to tackle climate change, I soon realised that they are fantastic tools for this job that are being overlooked. Overall, it seems that very few tools are being used for the job, which basically means everyone is either bumbling around trying to do their best, or not having any idea where to start.

This chapter gives you detail on the 'how to' of implementing the improvements in this book successfully, using Lean/Agile and some other techniques. This is crucial because we have to make it easy for ourselves to take the hugely important first steps and then remain motivated to carry on.

Lean and Agile

Here's a brief further explanation of Lean and Agile to understand how they can help us when we implement improvements and change.

I have used the terms of Lean and Agile interchangeably in the book. This is because both terms encapsulate that it's better to get going on something and make improvements *bit by bit*, rather than getting stuck before you even start.

When the problem seems too big and daunting to tackle, getting 'stuck' is really common. Or, you get caught up trying to imagine the perfect solution (and planning it in detail), not realising the futility of doing so, because details will always be different in the real world.

But it's important to note that there *are* differences between Lean and Agile, and we can use them in different ways to facilitate our success when making any changes.

Agile is a set of guiding principles that derive from the 'Lean' methodology. In turn, 'Scrum' is a framework that can be used to *apply* Agile principles when problem-solving. When people refer to Agile, they probably mean Scrum.

One of the key elements of Scrum is 'sprinting', which is a way to tackle problems by writing down all the individual activities you need to do to make an improvement come to fruition, then doing a chunk of them in each 'sprint'. A sprint is just a block of time, like a week or two-week period, with the goal of completing one increment of the improvement at the end of that period.

Continuous improvement is terminology from 'Lean'. It is one of the pillars of the methodology, and it refers to always striving to improve on the current situation with small changes that bring you closer and closer to the ideal. It is continual, consistent change.

So one approach is about implementing changes in defined blocks of time; the other is about continual incremental change. In business, they both have been proven to offer quicker, more effective, and longer-lasting change than big overhauls.

We can choose to use either Lean or Agile (Scrum) *or both of them* for different types of solutions. For example:

- **Continuous improvement** is a mindset change that you can adopt to support many improvements. We should apply this mindset of continual, consistent change when adapting our lives to meet our carbon footprint and wellbeing goals.
- **Sprinting**, on the other hand, can put some structure and order around the activities needed to make the improvements. And help us organise ourselves to implement them successfully.

Goal setting

Before we start to use Continuous Improvement or Scrum-style 'sprints', or any other technique to make an improvement, it helps to set the specific goal. We need to know where we are trying to get to. Then, we'll know if we're being successful and can prioritise the activities we need to take to get there.

We know what our overarching goals are (as outlined in chapter 2). Namely, to reduce our carbon footprint to 3 tonnes per year and improve our wellbeing. With those in mind, choose which improvement you're going to implement and give yourself a specific goal.

It's good practice to make your goals SMART – **S**pecific, **M**easurable, **A**chievable, **R**ealistic and **T**imely. For example, with buying new clothes, depending on your personal situation, your goal could be to:

Always aim to buy second-hand by checking for alternative options before looking for new clothes and – within two years – reducing your purchases of new clothes by 90%, from 50 new items per year, for example, to five.

If this type of thing is new to you and seems a bit formal, don't get caught up on it. The idea is just for you to have a think about what you do now, in relation to the lifestyle topics in the book, and have an idea of where you want to get to and when.

Continuous improvement

We've previously covered some of the benefits of Continuous Improvement. And we know that we need to manage some of the unhelpful traits of our animalistic brain if we want to overcome the CC22. So it's worth exploring more how Continuous Improvement can help us to do this too.

The lesser-developed parts of our brain include the Amygdala which houses the body's alarm mechanism or fear response. It is triggered by change and makes us more likely to resist it.

"We are built to resist radical change… our nervous system wires us for resistance to a big overhaul of any kind," writes Bob Maurer in 'The Spirit of Kaizen'.

With that in mind, when you implement any improvements, make sure you aim to change one very small thing at a time. It has to be a change that is minor enough to avoid or bypass the brain's fear response so that it can be done with ease. If we take our new clothes example, from above, there are several ways that we can gain a 2-tonne carbon footprint saving. A fundamental one is to cut down the new clothes you buy *by a small amount at a time*, so you don't get scared off at the thought of changing your habits overnight.

What that amount is will be personal to you; it could be buying one less item a week, or a month, for example. Whatever it is, take the first little step, followed by the next and the next. Once you're up and running, you'll find the more you do it, the easier it becomes. Soon, you'll have built new neural pathways in your brain which then snowball into new behaviours that you don't think twice about.

I can 100% personally vouch for this. I used to pop into shops regularly and buy things on a whim without giving it any thought at all. At the time, if someone had said to me that I needed to stop tomorrow, I'd have been pretty horrified. But, by adopting this approach, it's now completely normal for me to avoid the clothes shops. More than that, I look back and can barely believe how much time I spent in stores, considering I never found them remotely fulfilling and it was all just habit reinforced by the CC22 effect.

While implementing a continuous improvement approach, you can use other techniques, which will keep you motivated and make the whole thing completely palatable. See the section entitled 'Motivation' later in the chapter for more.

If you're interested in the topic, you can read more about how to apply a Continuous Improvement mindset to your life to make positive lasting change. The Bob Maurer quote (above) is taken from his book, which is a brilliant place to start.[162]

Scrum - Sprinting

As mentioned, a 'sprint' is a block of time within which you complete a chunk of activities. But, firstly, you need to write all the activities down and prioritise them.

Prioritisation

I'm sure that many of us already use 'to do' lists, but how many of us prioritise them? It's not something we normally do because it's hard to prioritise *anything* without thinking about what we're trying to achieve first. So, have your specific goal in mind as you prioritise, to make it easier.

Write down everything that you'll need to do to achieve your goal and then number the activities in order of precedence. It may be that some of the things you've written down are more of a 'nice to have' than an essential element of getting the improvement done; these things go at the bottom.

Then, you're going to take a chunk of the high-priority activities that you feel you can complete in the first Sprint. How long the Sprint will last is up to you, but normally it's a week or two weeks. Don't make it longer than a month, or you'll risk losing focus. It's important to complete a list of activities in the Sprint that will give you something tangible at the end. In the corporate world, this is so you can get feedback from colleagues or customers on what you've managed to achieve, but in this case, it may be something that you share with a friend or loved one instead.

Let's apply this framework, so far, to an example in the book and bring things to life. In the chapter on consumer choices, there's a potential yearly carbon footprint saving of 0.8 tonnes if you don't buy new household furniture regularly. One of the improvements you can implement in this category is to replace buying new with restoring old furniture instead.

Let's say you're looking for a set of dining table and chairs; you need to write down all the activities that you'd need to do to reach your goal which is to: have a fully-restored and ready-to-use table and chairs in your property in one month.

Below is a subset of what the activities could be, to give you an idea. I've already put them in priority order but just bear in mind that you may rank things a bit differently depending on what's more important to you, (e.g., you may want to shop in stores rather than online).

1. Search online for a second-hand dining set to buy.
2. Decide which set you want to buy.
3. Agree a price with the vendor.
4. Search online for restoration ideas on Pinterest.
5. Select three possible restoration options on Pinterest.
6. Decide on the restoration idea of choice.
7. Buy dining table set.
8. Arrange for delivery/pick up.
9. Search online for restoration services.
10. Call three restoration services for quotes.
11. Choose a quote.
12. Book dining table set in to be restored.
13. Search online for second-hand furniture stores in the area.
14. Visit second-hand stores in the area.

In the first Sprint, you might decide you can do numbers 1 to 5. Don't decide to stop at number 4, because you want something tangible at the end that you can showcase. In this instance, it would be the Pinterest pics that you show someone for a second opinion. Having something to show for your efforts – even if it's just to yourself – will help keep you motivated to crack on until the overall task is done.

If you don't complete all the tasks you thought you would in the Sprint, then *don't* add time onto the end. Add them into the next Sprint, instead. And if you finish early, *don't* add extra tasks into the Sprint; start a new one instead. This allows you to have a quick assessment at the end of each Sprint to gauge what has been particularly successful, and what tactics you might want to work on next time. For example, if you really struggled to do your internet searches in the evening – as you're too wiped out – then you might want to do them in your lunch hour next time instead.

Scrum summary

This is just a small part of the Scrum framework, and as you can see, it is highly-structured. You may be a mega-organised person who sails through implementing the improvements and stays focussed naturally, but if you struggle to get things done within a specific period (or full stop), then this approach takes very little time to do compared to the time it can save.

It's hard to stay on top of things when you're trying to remember all the things you need to do, off the top of your head, and in which order. Frameworks are meant to be flexible, so there may be parts of it that you find really helpful (and adopt), and parts that you feel don't help much (so you leave them out).

There is a lot more detail on the Scrum framework that is outside the scope of this book. Maybe you're keen to know more of the uses for it, such as how to find out whether you're on the right track of reaching your overall carbon footprint goal within the five-year timeframe. If so, then the book "Scrum: The Art of Doing Twice the Work in Half the Time" by Jeff Sutherland (who co-created it) is a great read.[163]

Other Lean 'tricks' - 'How to' with less fuss and no muss

The 5S System

Lean teaches 'The 5S System' in the workplace to ensure that you have the materials you need to do something in the right place, *when* you need them. It, therefore, helps reduce wasted time and effort.

Here are the 5S pillars with how they can be applied in relation to recycling, but you could also apply them to buying and storing food, or clothes, for example:

1. Sort – sort any equipment you need

- This will mainly be:

1. Boxes/bags to store your various types of recycling until they're collected, or you drop them off at the recycling facility.
2. Labels for each box.
3. A place where you can leave rinsed recycling to dry.

2. Straighten – straighten things up

- Label your boxes or other packaging clearly and have your equipment somewhere where it's not in your way, but where you have easy access. You'll be less likely to recycle if the recycling box is all the way down the hall, but the rubbish bin is right in front of you!

3. Scrub – keep it clean and maintained

- Rinse all recycling to avoid contamination. This has the added benefit that your storage place will be easy to maintain, as it'll be smell- and dirt-free, and you won't need any extra bin liners. Also, empty out and drop off recycling as regularly as you need to, to keep the storage area in order, so you don't have a messy overflow to look at.

4. Systemise – create a way of working

- The simpler, the better; we're not trying to do anything flashy here. It'll probably be something like:
  1. When packaging is empty, rinse and dry with other washing up and place in appropriate recycling box/bag.
  2. When full, take outside to recycling collection bin and put collection bin in the right place for collection on the appropriate day.

5. Standardise – stick to a system that works consistently, but review regularly for refinements

- Create your system with one type of recycling and build it up from there. Make sure everyone else in the home understands the system and take feedback on how to make it better.

'Poke Yokes'

In Lean, Poke Yokes (which is the Japanese term, as this was where the methodology was refined) is equivalent to error-proofing or fail-safes.

There are various types of fail-safes which are designed into processes or products to help us get things right. For example, when your car beeps at you when you don't put your seat belt on, or when the lights automatically turn off when you switch off the engine.

Fail-safes are normally simple and inexpensive, but they do the trick for preventing things from going wrong. We can use them to keep us on track, like with our recycling routine. For example, it's probably worth setting a reminder or alarm on your phone for when the various recycling collection/drop off days are, as some may only be every few weeks.

Motivation

There are other elements to success that can keep us on the right track in terms of making improvements happen; keeping motivated is a big one.

When using the following motivational techniques, alongside some of the tools and practices above, implementing change can be easy and rewarding instead of unmanageable and daunting.

Money and happiness

Many of the improvements in this book have the happy side-effect of saving you money. Yay!

However, it's not the size of the wad in our pocket, but what we do with it that counts. If we want money to bring us happiness, studies have shown that we're actually doing *ourselves* a favour when we spend our money on others instead of ourselves.[164] The money you will often save when implementing the improvements listed in this book could be put to good use by helping those who are in greater need of it. This will give you a wellbeing boost and, at the same time, will motivate you to keep doing more.

The other thing we can do with our cash – that will have a huge impact on our carbon footprint and improve our wellbeing (as we saw in chapter 8) – is to spend it on experiences rather than material things.

Using the method of scoring your wellbeing (recommended in chapter 2), *keep track* of your improvements. This means that when you check in on your wellbeing score after each one, you can reward yourself at milestones. This is a great motivator in itself, but to get maximum motivation and satisfaction, reward yourself with an experience. Experiences are far more memorable and enjoyable; get that massage or meal with friends booked in, and know that you thoroughly deserve it!

One thing at a time

The strategy of staying focussed on one thing at a time is a key Lean/Agile principle, and like many of the Lean/Agile principles, is cited in countless other books and studies as being a key to success.

Our brains lose focus when there are several things competing for attention. No matter how good at multitasking we think we might be, the neuroscience proves that we can only do one thing at a time competently. It is literally impossible to focus – to the best of our abilities – on more than one thing at a time.

This explains why there are so many more road accidents when people are using their phones. For this reason, you should focus on completing one improvement at a time for the quickest route to reach your carbon footprint saving goal. If you're using continuous improvement to reduce a particular habit or behaviour, and after a while it becomes second nature and no longer needs focus, you can begin another improvement.

The 'one thing at a time' mentality applies to the individual tasks within an improvement as well. The structure of sprinting can help because it can focus you on what particular tasks to complete and in what order. This is more motivating than trying to do a little bit of each activity simultaneously, and becoming overwhelmed by everything that you need to do.

Share the love

Another great way to stop you forgetting, or not getting around to implementing improvements, as well as motivating yourself along the way – is to *tell other people* what you're doing. And, wherever possible, get them involved.

Tell your friends, both in-person and on social media, what you're up to; this will not only make you more likely to follow through with something but will also evoke responses from them that help keep you focussed and feeling like you're doing the right thing.

Spreading the word is going to be a big element of the collective success of individual actions. The more people you tell, the more people will follow suit. Once momentum begins to build, cultural shifts can happen quickly, so we have a real chance of getting where we need to be.

Quick productivity tip

A top tip for being productive and achieving goals is to either take action right away, or plan a specific time that you will. So, for example, if you want to check your energy supplier's credentials or change supplier (chapter 6), put this book down and get online right now. Otherwise, visualise in your head when you'll do it, think about what you'll be doing before and after, and make sure you have a suitable time slot and a place in mind. Then, write down or say out loud that you're going to do it, and tell someone else that you're going to do it too. All this makes it much more likely that you will actually take action, rather than never quite getting around to it.

A lot is at stake

A big one that has really helped me when I've been feeling disheartened, or found something inconvenient, is reminding myself what is at stake. If you've been exposed to any of the information that scientists are telling us about what is coming, you know things are pretty dire. If you're reading this book, then you probably agree that we all need to take responsibility in order to resolve the problem. So, when you're not sure whether the tasks in hand are worth any inconvenience, have a think about:

1. *Why* you're doing this in the first place.
2. *Who* stands to benefit or suffer as a result of what each of us do now.
3. *Whether* making the effort will bring you satisfaction.

There may be a trade-off involved in some of the things you do. But becoming more of the person you want to be, and experiencing fulfilment as a result, can make it all worthwhile!

Endnotes

[1] *Climate Change.* United Nations. https://www.un.org/en/sections/issues-depth/climate-change/

[2] *The UK Climate Change Risk Assessment Evidence Report* (2017). Climate Change Committee. https://www.theccc.org.uk/uk-climate-change-risk-assessment-2017/

[3] *Only rebellion will prevent an ecological apocalypse.* The Guardian, https://www.theguardian.com/commentisfree/2019/apr/15/rebellion-prevent-ecological-apocalypse-civil-disobedience

[4] *The Deloitte Consumer Review – The growing power of consumers.* Deloitte. https://www2.deloitte.com/content/dam/Deloitte/uk/Documents/consumer-business/consumer-review-8-the-growing-power-of-consumers.pdf

[5] *Consumers have huge environmental impact.* Science Daily. https://www.sciencedaily.com/releases/2016/02/160224132923.htm

[6] https://www.brainyquote.com/quotes/margaret_mead_100502

[7] *The success of nonviolent civil resistance.* Erica Chenoweth. TEDxBoulder. https://youtu.be/YJSehRlU34w

[8] *The '3.5% rule': How a small minority can change the world.* BBC. https://www.bbc.com/future/article/20190513-it-only-takes-35-of-people-to-change-the-world

[9] *Climate Change in the American Mind.* Yale University; George Mason University. https://climatecommunication.yale.edu/wp-content/uploads/2019/01/Climate-Change-American-Mind-December-2018.pdf

[10] *Plant Based Food Products Started With Milk, Now Taking On Meat, What's Next?* Forbes. https://www.forbes.com/sites/bernhardschroeder/2019/06/18/plant-based-food-products-started-with-milk-now-taking-on-meat-whats-next/?sh=67879b021da8

[11] *Mitigation Pathways Compatible with 1.5°C in the Context of Sustainable Development.* The Intergovernmental Panel on Climate Change. https://www.ipcc.ch/site/assets/uploads/sites/2/2019/02/SR15_Chapter2_Low_Res.pdf

[12] *How do CO2 emissions compare when we adjust for trade?* Our World in Data. https://ourworldindata.org/consumption-based-co2

[13] CO2e stands for carbon dioxide and carbon dioxide equivalents which are the various greenhouse gasses that are contributing to climate change and are therefore included in your footprint. Where CO2 is referenced throughout the book it is normally referring to CO2e.

[14] *Low Carbon Solutions for a Sustainable Consumer Goods Sector.* The Consumer Goods Forum. https://www.theconsumergoodsforum.com/wp-content/uploads/2017/12/low-carbon-solutions-sustainable-consumer-goods.pdf

[15] *A healthy economy should be designed to thrive not grow.* Kate Raworth. Ted. https://www.ted.com/talks/kate_raworth_a_healthy_economy_should_be_designed_to_thrive_not_grow?language=en

[16] *Finding Brand Success In The Digital World.* Forbes. https://www.forbes.com/sites/forbesagencycouncil/2017/08/25/finding-brand-success-in-the-digital-world/#3f8674c5626e

[17] *Consumerism and its discontents.* American Psychological Association. https://www.apa.org/monitor/jun04/discontents

[18] *Innovation Loves a Crisis.* Psychology Today. https://www.psychologytoday.com/us/blog/the-tao-innovation/200904/innovation-loves-crisis

[19] *Origin of the concept 'circular fashion'.* Green Strategy. https://circularfashion.com/circular-fashion-definition/

[20] *Sustainability, Circularity.* Stella McCartney. https://www.stellamccartney.com/experience/en/sustainability/circularity-2/

[21] *Can Recycled Plastic Clothing do More Harm than Good?* Eluxe Magazine. https://eluxemagazine.com/magazine/recycled-plastic-clothing/

[22] *Ikea.* Peoples Assembly of Victoria. https://www.paov.ca/mediamenu/alternative-news/12613-ikea1596882121

[23] *Palm Oil.* Greenpeace. https://www.greenpeace.org.uk/what-we-do/forests/deforestation-climate-change/

[24] *7 of the Most Eco-Friendly Cell Phones on the Market.* Tree Hugger. https://www.treehugger.com/gadgets/most-eco-friendly-cell-phones-market.html

[25] Find out more at fairphone.com.

[26] *A healthy economy should be designed to thrive not grow.* Kate Raworth. Ted. https://www.ted.com/talks/kate_raworth_a_healthy_economy_should_be_designed_to_thrive_not_grow?language=en

[27] *The role of national culture in shaping public policy: a review of the literature. HC Coombs Policy Forum Discussion.* (2014). Research Gate. https://www.researchgate.net/publication/266273390_The_role_of_nation al_culture_in_shaping_public_policy_a_review_of_the_literature

[28] *The Earth is Full.* Paul Gilding. Ted. https://www.ted.com/talks/paul_gilding_the_earth_is_full?language=en

[29] *How the oil industry has spent billions to control the climate change conversation.* The Guardian. https://www.theguardian.com/business/2020/jan/08/oil-companies-climate-crisis-pr-spending

[30] *Study: Volkswagen's excess emissions will lead to 1,200 premature deaths in Europe.* MIT News. http://news.mit.edu/2017/volkswagen-emissions-premature-deaths-europe-0303

[31] *A global snapshot of the air pollution-related health impacts of transportation sector emissions in 2010 and 2015.* The International Council on Clean Transportation (ICCT).

https://theicct.org/publications/health-impacts-transport-emissions-2010-2015

[32] *A global snapshot of the air pollution-related health impacts of transportation sector emissions in 2010 and 2015.* The International Council on Clean Transportation (ICCT). https://theicct.org/publications/health-impacts-transport-emissions-2010-2015

[33] *Report: A Resilient Future for Coastal Communities Federal Policy Recommendations from Solutions in Practice.* Environmental and Energy Study Institute. https://www.eesi.org/papers/view/fact-sheet-fossil-fuel-subsidies-a-closer-look-at-tax-breaks-and-societal%20costs

[34] *Reforming Subsidies Could Help Pay for a Clean Energy Revolution: Report.* Global Subsidies Initiative. https://www.iisd.org/gsi/news-events/reforming-subsidies-could-help-pay-clean-energy-revolution-report

[35] This was worked out using worldwide data for the number of hours flown by the average flyer, and UK data for the average miles driven and fuel consumption. That means this figure is considerably more for drivers in the US and varies from country to country.

[36] *Around the World in 50 hours: Average person takes 6.5 flights per year.* News 24. https://www.news24.com/news24/travel/around-the-world-in-50-hours-average-person-takes-65-flights-per-year-20180828

[37] I appreciate that I'm using worldwide data for flying and UK data for driving here but we don't need to get bogged down in that. For the purposes of ordering the improvements, the data serves its purpose and gives us the indication we need to get going.

[38] *Transport.* The Intergovernmental Panel on Climate Change. https://www.ipcc.ch/site/assets/uploads/2018/02/ipcc_wg3_ar5_chapter8.pdf

Airlines' CO2 emissions rising up to 70% faster than predicted. The Guardian. https://www.theguardian.com/business/2019/sep/19/airlines-co2-emissions-rising-up-to-70-faster-than-predicted

Climate explained: how much does flying contribute to climate change? The Conversation. https://theconversation.com/climate-explained-how-much-does-flying-contribute-to-climate-change-127707

[39] *This Is Your Brain on Nature.* National Geographic Magazine. https://www.nationalgeographic.com/magazine/2016/01/call-to-wild/

[40] *Responsible Tourism.* Responsible Travel. https://www.responsibletravel.com/holidays/responsible-tourism/travel-guide

[41] *Can 5G really be sustainable?* Raconteur. https://www.raconteur.net/sustainability/sustainable-business-2020/5g-environmental-impact

[42] *How long does it take a rainforest to regenerate?* New Scientist. https://www.newscientist.com/article/dn14112-how-long-does-it-take-a-rainforest-to-regenerate/#ixzz6PYlgS8LB

[43] *How trees talk to each other.* Suzanne Simard. Ted. https://www.ted.com/talks/suzanne_simard_how_trees_talk_to_each_other/transcript?language=en

[44] *Vehicle mileage and occupancy.* Gov.UK. https://www.gov.uk/government/statistical-data-sets/nts09-vehicle-mileage-and-occupancy#car-mileage

Average MPG for Cars UK 2020. Nimble Fins. https://www.nimblefins.co.uk/average-mpg

Energy and environment: data tables (ENV). Gov.UK. https://www.gov.uk/government/statistical-data-sets/energy-and-environment-data-tables-env#fuel-consumption-env01

[45] Waze. https://www.waze.com/en-GB/carpool/companies?city=Mountain%20View

[46] *Driving too close caused circa 1 in 8 casualties in 2016, including 114 killed or seriously injured.* (Analysis by Highways England). Reported in the Sunday Times. https://www.driving.co.uk/news/drive-close-car-front-risk-100-fine-new-clampdown-tailgaters/

[47] *Life Cycle Analysis of the Climate Impact of Electric Vehicles.*
Transport and Environment.
https://www.transportenvironment.org/sites/te/files/publications/TE%20-
%20draft%20report%20v04.pdf

[48] *Net emission reductions from electric cars and heat pumps in 59 world regions over time.* Nature. https://www.nature.com/articles/s41893-020-0488-7#citeas

[49] Mitsubishi. https://www.mitsubishi-motors.com/en/innovation/motorshow/2019/gms2019/dendo/

[50] *OVO Vehicle-to-Grid Trial.* OVO Energy.
https://www.ovoenergy.com/electric-cars/vehicle-to-grid-charger

[51] *Tesla stock soars, worth more than Big Oil, as Battery Day approaches.*
The Driven. https://thedriven.io/2020/08/24/tesla-stock-soars-worth-more-than-big-oil-as-battery-day-approaches/

[52] *Electric Cars.* Ethical Consumer.
https://www.ethicalconsumer.org/transport-travel/shopping-guide/electric-cars

[53] *Airlines' CO2 emissions rising up to 70% faster than predicted.* The
Guardian. https://www.theguardian.com/business/2019/sep/19/airlines-co2-emissions-rising-up-to-70-faster-than-predicted

[54] "Emissions per passenger kilometre have been reduced by more than 50% since 1990. According to the airline trade body IATA… An IATA spokesman said, "It is true that because of demand from people in developing economies to enjoy the same benefits of flying as those in rich countries, aviation emissions growth is currently faster than our efficiency gains." *Airlines' CO2 emissions rising up to 70% faster than predicted.*
The Guardian.
https://www.theguardian.com/business/2019/sep/19/airlines-co2-emissions-rising-up-to-70-faster-than-predicted

[55] *How much of a cancer risk is processed meat?* BBC.
https://www.bbc.co.uk/food/articles/processed_meat_danger

[56] *The China Study* by Colin Campbell; pages 231-232

[57] *Feeding the future: Fixing the world's faulty food system.* The Telegraph. https://www.telegraph.co.uk/news/feeding-the-future/

[58] *Cows.* Compassion in World Farming. https://www.ciwf.org.uk/farm-animals/cows/

[59] *The Secrets of Food Marketing.* Kate Cooper. YouTube. https://www.youtube.com/watch?v=mKTORFmMycQ

[60] *Cognitive Dissonance, Willpower, and Your Brain.* Psychology Today. https://www.psychologytoday.com/gb/blog/what-would-aristotle-do/201809/cognitive-dissonance-willpower-and-your-brain

[61] *Cognitive Dissonance, Willpower, and Your Brain.* Psychology Today. https://www.psychologytoday.com/gb/blog/what-would-aristotle-do/201809/cognitive-dissonance-willpower-and-your-brain

What Happens to the Brain During Cognitive Dissonance? Scientific American. https://www.scientificamerican.com/article/what-happens-to-the-brain-during-cognitive-dissonance1/

[62] *Key facts and findings.* Food and Agriculture Organization of the United Nations. http://www.fao.org/news/story/en/item/197623/icode/

[63] This compares to the direct emissions from transport, i.e., the emissions released directly from transport but not emissions produced during production or disposal of vehicles.

[64] *Feeding the future: Fixing the world's faulty food system.* The Telegraph. https://www.telegraph.co.uk/news/feeding-the-future/

[65] *Livestock's Long Shadow: environmental issues and options.* Food and Agriculture Organization of the United Nations. http://www.fao.org/3/a0701e/a0701e.pdf

[66] *Air Pollution from Agriculture.* Department for Environment, Food & Rural Affairs. https://uk-air.defra.gov.uk/assets/documents/reports/aqeg/2800829_Agricultural_emissions_vfinal2.pdf

[67] *Air Pollution from Agriculture.*

Department for Environment, Food & Rural Affairs. https://uk-air.defra.gov.uk/assets/documents/reports/aqeg/2800829_Agricultural_em issions_vfinal2.pdf

[68] *Understanding Cognitive Dissonance (and Why it Occurs in Most People).* Cleverism. https://www.cleverism.com/understanding-cognitive-dissonance-and-why-it-occurs-in-most-people/

[69] Beef = almost 300 kg CO2-eq per kilogram of protein produced, meat and milk from small ruminants = 165 and 112kg CO2-eq.kg respectively. Cow milk, chicken products and pork = below 100 CO2-eq/kg. From country to country, within each commodity type there is very high variability in emission intensities due to the different practices/inputs to production used. *Key facts and findings.* Food and Agriculture Organization of the United Nations. http://www.fao.org/news/story/en/item/197623/icode/

[70] *Cars or livestock: which contribute more to climate change?* Thomson Reuters Foundation News. http://news.trust.org/item/20180918083629-d2wf0

[71] *How Much Food Do We Waste? Probably More Than You Think.* The New York Times. The New York Times. https://www.nytimes.com/2017/12/12/climate/food-waste-emissions.html

[72] *Why Save Food?* Love Food Hate Waste. https://lovefoodhatewaste.com/why-save-food

[73] *Food wastage footprint & Climate Change.* Food and Agriculture Organization of the United Nations. http://www.fao.org/3/a-bb144e.pdf

[74] *Greenhouse Gas Emissions from Food Loss and Waste Approach the Levels from Road Transport.* World Resources Institute. https://wriorg.s3.amazonaws.com/s3fs-public/uploads/FLW_graphic1.jpg

[75] *A-Z of Food Storage.* Love Food Hate Waste. https://lovefoodhatewaste.com/article/food-storage-a-z

[76] *All These Jobs Would Be Lost If the UK Went Completely Vegan.* Vice. https://www.vice.com/en/article/59dpdk/all-these-jobs-would-be-lost-if-the-uk-went-completely-vegan

[77] *How is climate change affecting the economy and society?* Iberdrola. https://www.iberdrola.com/environment/impacts-of-climate-change

[78] *UK's hottest recorded day 'caused deaths of extra 200 people'.* The Guardian. https://www.theguardian.com/world/2019/dec/11/uks-hottest-recorded-day-caused-deaths-of-extra-200-people

[79] *Heat and health.* European Environment Agency. https://www.eea.europa.eu/data-and-maps/indicators/heat-and-health/heat-and-health-assessment-published

[80] *Consumers have huge environmental impact.* Science Daily. https://www.sciencedaily.com/releases/2016/02/160224132923.htm

[81] *A Fossil Fuel-Free Industrial Revolution.* Advanced Science News. https://www.advancedsciencenews.com/would-a-fossil-fuel-free-industrial-revolution-have-been-possible/

[82] *Is green growth possible?* By Jason Hickel and Giorgos Kallis. New Political Economy Journal. https://www.tandfonline.com/doi/full/10.1080/13563467.2019.1598964

[83] *The GENI Initiative and 100% Renewable Energy Reports.* Global Energy Network Institute. http://www.geni.org/globalenergy/research/100-percent-renewable-energy-reports/index.shtml

[84] *100% Clean and Renewable Wind, Water, and Sunlight All-Sector Energy Roadmaps for 139 Countries of the World.* Joule. https://web.stanford.edu/group/efmh/jacobson/Articles/I/CountriesWWS.pdf

[85] *How the oil industry made us doubt climate change.* BBC News. https://www.bbc.co.uk/news/stories-53640382. For the record, Exxon claimed that the accusations are "baseless and without merit".

[86] *Election 2019: What the manifestos say on energy and climate change.* Carbon Brief. https://www.carbonbrief.org/election-2019-what-the-manifestos-say-on-energy-and-climate-change

[87] *BP's statement on reaching net zero by 2050 – what it says and what it means.* The Guardian. https://www.theguardian.com/environment/ng-

interactive/2020/feb/12/bp-statement-on-reaching-net-zero-carbon-emissions-by-2050-what-it-says-and-what-it-means

[88] *Election 2019: What the manifestos say on energy and climate change.* Carbon Brief. https://www.carbonbrief.org/election-2019-what-the-manifestos-say-on-energy-and-climate-change

[89] *Fossil fuels and climate change: the facts.* Client Earth. https://www.clientearth.org/fossil-fuels-and-climate-change-the-facts/

[90] *Net Zero: The UK's contribution to stopping global warming.* Committee on Climate Change. May 2019. https://www.theccc.org.uk/wp-content/uploads/2019/05/Net-Zero-The-UKs-contribution-to-stopping-global-warming.pdf

[91] *Tracing Fossil Fuel Companies' Contributions to Temperature Increase and Sea Level Rise.* Union of Concerned Scientists Factsheet. https://www.ucsusa.org/sites/default/files/attach/2017/10/gw-accountability-factsheet.pdf

[92] When calculating the potential carbon footprint saving for this chapter, I used a medium-sized home. In the chapters on food and consumer products, I quoted the maximum potential carbon saving possible. The reason I've done this is because a lot of people are eating meat and dairy with most meals, and also buying 'stuff' like clothes and gadgets regularly. However, in relation to housing, the amount of carbon used to power a house rises tremendously depending on size. That makes it a bit silly to quote the largest possible carbon saving, unless you happen to be a multi-millionaire or billionaire with a 15-bedroom mansion for a home. I'm assuming most people reading this don't.

[93] *How Green is Green Gas?* Physics World. https://physicsworld.com/a/how-green-is-green-gas/

[94] *Revolving Door.* Investopedia. https://www.investopedia.com/terms/r/revolving-door.asp

[95] *What is green gas?* Ecotricity. https://www.ecotricity.co.uk/our-green-energy/our-green-gas/what-is-green-gas

[96] *The Health Impacts of Cold Homes and Fuel Poverty.* Marmot review team. http://www.instituteofhealthequity.org/resources-reports/the-health-

impacts-of-cold-homes-and-fuel-poverty/the-health-impacts-of-cold-homes-and-fuel-poverty.pdf

⁹⁷ *Does demolition or refurbishment of old and inefficient homes help to increase our environmental, social and economic viability?* Energy Policy.
https://www.london.gov.uk/sites/default/files/ad_52_anne_power_-_does_demolition_or_refurbishment_.pdf

⁹⁸ *Light Bulbs.* Ethical Consumer.
https://www.ethicalconsumer.org/energy/shopping-guide/light-bulbs

⁹⁹ *How to Insulate your Home.* WikiHow.
https://www.wikihow.com/Insulate-Your-Home

¹⁰⁰ *Roof and loft insulation.* Energy Saving Trust.
https://energysavingtrust.org.uk/home-insulation/roof-and-loft

¹⁰¹ *Roof and loft insulation.* Energy Saving Trust.
https://energysavingtrust.org.uk/home-insulation/roof-and-loft

¹⁰² *Regeneration and Retrofit.* UK GBC Task Group Report.
https://www.ukgbc.org/wp-content/uploads/2017/09/08498-Regen-Retrofit-Report-WEB-Spreads.pdf

¹⁰³ *Regeneration and Retrofit.* UK GBC Task Group Report.
https://www.ukgbc.org/wp-content/uploads/2017/09/08498-Regen-Retrofit-Report-WEB-Spreads.pdf

¹⁰⁴ *Quick tips to save energy.* Energy Saving Trust.
https://energysavingtrust.org.uk/home-energy-efficiency/energy-saving-quick-wins

¹⁰⁵ *How much energy could be saved by making small changes to everyday household behaviours?* Department of Energy &Climate Change.
https://assets.publishing.service.gov.uk/government/uploads/system/uploads/attachment_data/file/128720/6923-how-much-energy-could-be-saved-by-making-small-cha.pdf

¹⁰⁶ *Regeneration and Retrofit.* UK GBC Task Group Report.
https://www.ukgbc.org/wp-content/uploads/2017/09/08498-Regen-Retrofit-Report-WEB-Spreads.pdf

[107] *How do Heat Recovery Systems Work?* BPC Ventilation. https://www.bpcventilation.com/blog/do-heat-recovery-systems-work

[108] *A guide to how infrared heating systems work.* Silver Surfers. https://www.silversurfers.com/financial/energy/guide-infrared-heating-systems-work/

[109] *Regeneration and Retrofit.* UK GBC Task Group Report. https://www.ukgbc.org/wp-content/uploads/2017/09/08498-Regen-Retrofit-Report-WEB-Spreads.pdf

[110] *Retrofitting British homes to make them more energy efficient.* Climate Xchange. https://www.climatexchange.org.uk/blog/retrofitting-british-homes-to-make-them-more-energy-efficient/

[111] *Net Zero: The UK's contribution to stopping global warming.* Committee on Climate Change. May 2019. https://www.theccc.org.uk/wp-content/uploads/2019/05/Net-Zero-The-UKs-contribution-to-stopping-global-warming.pdf

[112] *Recycling Statistics and Facts.* All-Recycling-Facts. http://www.all-recycling-facts.com/recycling-statistics.html

[113] *Seventh Resource.* Global Recycling Foundation. https://www.globalrecyclingday.com/seventh-resource/

[114] *Is what we're recycling actually getting recycled?* How stuff works. https://science.howstuffworks.com/environmental/conservation/issues/recycling-reality.htm

[115] *Evaluation of landfill gas emissions from municipal solid waste landfills for the life-cycle analysis of waste-to-energy pathways.* Science Direct. https://www.sciencedirect.com/science/article/pii/S0959652617317316

[116] *Ecobricking secures plastic out of the biosphere, generates brikcoins and creates reusable building blocks for regenerative building.* GoBrick. https://www.gobrik.com/#

[117] *How Much Food Do We Waste? Probably More Than You Think.* The New York Times. https://www.nytimes.com/2017/12/12/climate/food-waste-emissions.html

[118] *How is Waste Food Recycled?* Recycle Now. https://www.recyclenow.com/recycling-knowledge/how-is-it-recycled/food-waste

[119] *The Challenge.* UNECE Sustainable Development Goals. https://www.unece.org/energywelcome/areas-of-work/methane-management/the-challenge.html

[120] *Is what we're recycling actually getting recycled?* How stuff works. https://science.howstuffworks.com/environmental/conservation/issues/recycling-reality1.htm

[121] *Fun recycling facts for children.* Business Waste. https://www.businesswaste.co.uk/recycling/fun-recycling-facts-for-children/

[122] *Fun recycling facts for children.* Business Waste. https://www.businesswaste.co.uk/recycling/fun-recycling-facts-for-children/

[123] *Recycling Statistics and Facts.* All-Recycling-Facts. http://www.all-recycling-facts.com/recycling-statistics.html

[124] *Services in Bournemouth.* BCP Council. https://www.bournemouth.gov.uk/BinsRecycling/RecyclingRecyclingCentres/AtoZofRecycling.aspx

[125] *Environment.* Little Green Paint & Paper. https://www.littlegreene.com/greene-standard

[126] *Cleaning products as bad for lungs as smoking 20 cigarettes a day, scientists warn.* Independent. https://www.independent.co.uk/news/health/cleaning-products-lungs-damage-cigarettes-smoking-20-day-scientists-warning-a8214051.html

[127] *Not so pretty: women apply an average of 168 chemicals every day.* The Guardian. https://www.theguardian.com/lifeandstyle/2015/apr/30/fda-cosmetics-health-nih-epa-environmental-working-group

[128] *Consumers have huge environmental impact.* Science Daily. www.sciencedaily.com/releases/2016/02/160224132923.htm

[129] *It feels good to do good.* Who Gives a Crap. https://uk.whogivesacrap.org/pages/our-impact

[130] *Streaming shows online could be damaging the environment.* Wired. https://www.wired.co.uk/article/amazon-apple-facebook-netflix-renewable-energy-greenpeace

[131] *With 'thank you' emails, polite Britons burn thousands of tonnes of carbon a year.* Reuters. https://www.reuters.com/article/us-britain-climate-change-tech-trfn/with-thank-you-emails-polite-britons-burn-thousands-of-tonnes-of-carbon-a-year-idUSKBN1Y029Q

[132] *Ecosia - The search engine that plants trees.* Chrome web store. https://chrome.google.com/webstore/detail/ecosia-the-search-engine/eedlgdlajadkbbjoobobefphmfkcchfk

[133] *Tourism responsible for 8% of global greenhouse gas emissions, study finds.* Carbon Brief. https://www.carbonbrief.org/tourism-responsible-for-8-of-global-greenhouse-gas-emissions-study-finds

[134] *The Quest for Adventure.* Barefoot. https://gobarefoot.travel/ecotourism-adventures/

[135] *What can you do to travel responsibly?* GSTC. https://www.gstcouncil.org/for-travelers/

[136] *Luxury cruise giant emits 10 times more air pollution (SOx) than all of Europe's cars – study.* Transport and Environment. https://www.transportenvironment.org/press/luxury-cruise-giant-emits-10-times-more-air-pollution-sox-all-europe%E2%80%99s-cars-%E2%80%93-study

[137] *The world's largest cruise ship and its supersized pollution problem.* The Guardian. https://www.theguardian.com/environment/2016/may/21/the-worlds-largest-cruise-ship-and-its-supersized-pollution-problem

[138] *Feeding the largest cruise ships in the world.* CNN Travel. https://edition.cnn.com/travel/article/cruise-ships-food-supplies/index.html

[139] *Our commitments.* Ponant. https://uk.ponant.com/sustainable-development

[140] *Waste away: exploring sustainability initiatives on-board cruise ships.* Ship Technology. https://www.ship-technology.com/features/sustainable-cruise-ships/

[141] *Sustainable Flying: is sustainable air travel possible?* FlyGreen. https://flygrn.com/page/sustainable-air-travel

[142] *6 Ways to Fly More Sustainably.* Conde Nast Traveller. https://www.cntraveller.com/article/how-to-fly-more-sustainably

[143] *How to reduce your carbon footprint when you fly.* BBC News. https://www.bbc.co.uk/news/av/science-environment-48206946/how-to-reduce-your-carbon-footprint-when-you-fly

[144] *Flight-Free Travel.* The Telegraph. https://www.telegraph.co.uk/travel/flight-free-travel/

[145] *Giant Strides.* The Telegraph. https://www.theguardian.com/society/2006/may/17/guardiansocietysupplement4

[146] *Scotland to China and back again ... Cods 10,000-mile trip to your table.* The Herald. https://www.heraldscotland.com/news/12765981.scotland-to-china-and-back-again-cods-10000-mile-trip-to-your-table/

[147] *Why do countries import and export the same good?* Economics Help. https://www.economicshelp.org/blog/150455/agriculture/why-do-countries-import-and-export-the-same-good/

[148] *The Economics of Happiness.* Helena Norberg-Hodge. Tedx Talk, YouTube. https://youtu.be/4r06_F2FIKM

[149] *Switch: how to change things when change is hard* by Chip Heath and Dan Heath

[150] *Escaping Growth Dependency (Positive Money report launch, 18th January 2018)*. YouTube. https://youtu.be/AmslDULmq78

[151] *World's richest 0.1% have boosted their wealth by as much as poorest half.* The Guardian. https://www.theguardian.com/inequality/2017/dec/14/world-richest-increased-wealth-same-amount-as-poorest-half

[152] *Escaping Growth Dependency (Positive Money report launch, 18th January 2018)*. YouTube. https://youtu.be/AmslDULmq78

[153] *The NS Interview: Noam Chomsky.* New Statesman. https://www.newstatesman.com/international-politics/2010/09/war-crimes-interview-obama

[154] *What Is the Chimp Model?* Chimp Management. https://chimpmanagement.com/the-chimp-model/

[155] *Economic De-Growth for Ecological Sustainability and Social Equity.* Degrowth Conference Proceedings. https://degrowth.org/wp-content/uploads/2011/07/Degrowth-Conference-Proceedings.pdf

[156] *What is post-growth economics, and why is it necessary?* Post Growth Institute. https://www.postgrowth.org/about-post-growth-economics

[157] *The Venus Project, beyond politics, poverty and war.* The Venus Project. https://www.thevenusproject.com/

[158] *Is green growth possible? By Jason Hickel and Giorgos Kallis.* New Political Economy Journal. https://www.tandfonline.com/doi/full/10.1080/13563467.2019.1598964

[159] *Plan B – is there an alternative to economic growth?* Miklós Antal. TEDxDanubia 2014. Tedx Talk, YouTube. https://youtu.be/J9_Xc9wxByM

[160] *Renewable Energy Job Boom Creates Economic Opportunity As Coal Industry Slumps.* Forbes. https://www.forbes.com/sites/energyinnovation/2019/04/22/renewable-energy-job-boom-creating-economic-opportunity-as-coal-industry-slumps/#1356a4e43665

[161] *Nonviolent resistance proves potent weapon.* The Harvard Gazette. https://news.harvard.edu/gazette/story/2019/02/why-nonviolent-resistance-beats-violent-force-in-effecting-social-political-change/

[162] *The Spirit of Kaizen: Creating lasting excellence one small step at a time* by Bob Maurer. *Kaizen* is the Japanese term for continuous improvement, courtesy of Toyota who developed the concept.

[163] For anyone that knows a bit about Agile principles and the Scrum framework, you'll probably know about it in terms of software development, which is where it is commonly implemented. However, as the co-originator of Scrum – Jeff Sutherland – explains in his book on the subject, the framework has much wider use. It's now being used to solve a myriad of problems, including making education more productive in classrooms! Lean was thought up in WWII, and gained recognition following its cultivation by Toyota. It was originally used in earnest in manufacturing industries, but is now applied across a wide number of industries and sectors to solve problems.

[164] *Think Small: The Surprisingly Simple Ways to Reach Big Goals* by Owain Service and Rory Gallagher.

Other books you might like from The Publisher

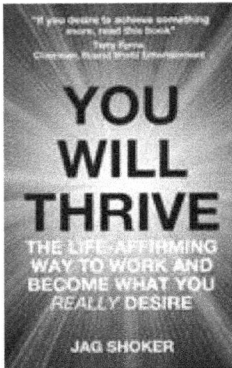

You Will Thrive: The Life-Affirming Way to Work and Become What You Really Desire

You Will Thrive addresses the subject of modern disillusionment. It is essential reading for people looking to make the most of their talents and be something more in life. Something that matters. Something that makes a difference in the world. Through six empowering steps, it reveals 'The Way' to boldly follow your heart as it leads you to the perfect opportunities you seek.

The 15-Minute Rule for Forgiveness

Forgiveness is one of the most powerful and liberating actions a person can take. Whether it is forgiving others, or oneself – for past deeds or mistakes – forgiveness can open people up to a life of happiness, fulfilment, and newfound accomplishment. The 15-Minute Rule is all about creating a safe framework for fostering forgiveness and self-forgiveness. We can all find 15 minutes in our busy lives and, through the short exercises and examples in the book, forgiveness and mental serenity can be attained.

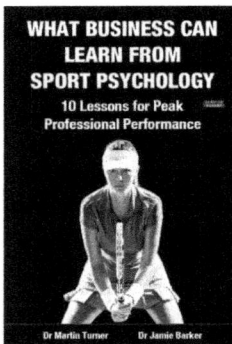

What Business Can Learn From Sport Psychology

The mental side of performance has always been a crucial component for success - but nowadays coaches, teams, and athletes of all levels and abilities are using sport psychology to help fulfil their potential and serve up success. It goes without saying that business performance has many parallels with sporting performance. Performance - in any context - is about utilizing and deploying every possible resource to fulfil one's potential. This book is about getting into a winning state of body and mind for performance.

www.ingramcontent.com/pod-product-compliance
Lightning Source LLC
Chambersburg PA
CBHW071547200326
41519CB00021BB/6644